Roseanne Santos de Carvalho
Gregorio G. Faccioli
Luciana C. Mendonça

Influence of wastewater reuse on sunflower microbiology

Roseanne Santos de Carvalho
Gregorio G. Faccioli
Luciana C. Mendonça

Influence of wastewater reuse on sunflower microbiology

Intended for animal feed

ScienciaScripts

Imprint

Cover image: www.ingimage.com

This book is a translation from the original published under ISBN 978-613-9-64330-1.

Publisher:
Sciencia Scripts
is a trademark of
Dodo Books Indian Ocean Ltd. and OmniScriptum S.R.L publishing group

120 High Road, East Finchley, London, N2 9ED, United Kingdom
Str. Armeneasca 28/1, office 1, Chisinau MD-2012, Republic of Moldova, Europe
Printed at: see last page
ISBN: 978-620-7-75841-8

SUMMARY

The use of treated wastewater for agricultural purposes can become an alternative for maintaining the quality of water bodies, the natural biota of the systems as well as relieving demand and preserving water supply for more restrictive uses. In addition to the aforementioned benefits, agricultural reuse has a positive impact on productivity and significant savings in chemical fertilisers. However, even though there are several advantages inherent in the use of reused water in agriculture, the presence of pathogens and organic contaminants must be highlighted. The aim of this study was therefore to assess the effects of reusing wastewater on the microbiological quality of sunflower crops (Helianthus annuus L.). The experiment was carried out in a greenhouse at the Department of Agronomic Engineering (DEA), located at the Federal University of Sergipe, São Cristóvão, between July and September 2012. The treated wastewater was collected from the Rosa Elze Sewage Treatment Plant (ETE), located in the municipality of São Cristóvão/SE. Irrigation was carried out using the following treatments: T1 (100% drinking water from the Sergipe Sanitation Company - DESO); T2 (100% treated wastewater); T3 (50% DESO water + 50% treated wastewater); T4 (25% DESO water + 75% treated wastewater) and T5 (75% DESO water + 25% treated wastewater). The irrigation rate was obtained using the FAO 56 method. A completely randomised design (DIC) was used, with 5 treatments and 4 replications per useful plot. The data obtained was analysed according to the parameters recommended by Resolution no. 12 of 02/01/2001 by the National Health Surveillance Agency - ANVISA. Parameters relating to climate, irrigation and water were assessed, but the main object of the study was the microbiological quality of the aerial part of the seedlings, which was enumerated for thermotolerant coliforms, E. coli, moulds and yeasts and Salmonella. The results obtained from the microbiological quality analyses show that the aerial part of the sunflower is within the standards established by current legislation, but further studies are needed, especially with regard to the soil.

KEYWORDS: Wastewater, Microbiological quality, Sunflower crop.

CONTENTS

CHAPTER 1

1. INTRODUCTION

Over the last 50 years, with the expansion of the urban population and the growth of industrial and technological development, the world's few available sources of fresh water are being jeopardised or put at serious risk. According to Rijsberman (2006), in the 20th century, the world's population tripled while water consumption increased sixfold. Various studies have concluded that two thirds of the world's population will be affected by water scarcity in the coming decades. The issue of water was the most relevant of all the proposals discussed at the Dialogues for Sustainable Development at the Rio +20 Conference (UN, 2012).

According to UNEP (2004), water scarcity is accompanied by a deterioration in water quality due to pollution and environmental degradation. The increase in the consumption of water supplies has led to a large increase in the volume of wastewater generated and, as a result, the addition of pollutants to natural waters. It is therefore necessary to raise awareness of the need to dispose of effluents in a safe way that benefits the entire planet. Effluent treatment is therefore a necessity for maintaining the quality of water bodies, the natural biota of the systems, as well as for conserving natural resources.

According to estimates, 72 per cent of the water used in Brazil is destined to meet the water demand of agricultural irrigation, with the remaining percentages going mainly to urban and industrial consumption. Therefore, sustainable development is one of the most important challenges for agriculture today, primarily because the term development itself is commonly confused with economic growth: the latter has a focus on increasing product consumption, while the term sustainable development has a bias towards preserving water resources, the environment, the practice of recycling and increased reuse.

According to Hespanhol (2003), planned water reuse is a potential alternative for rationalising this natural resource. The author stresses the importance of institutionalising, regulating and promoting water reuse in Brazil, ensuring that the practice is developed in accordance with appropriate technical principles, that it is economically viable, environmentally sustainable and socially acceptable and safe, in terms of environmental preservation and protection of the risk groups involved.

The benefits of reusing water from sewage treatment in agriculture are numerous. Studies carried out in various countries have shown that agricultural productivity increases significantly with the use of treated sewage. However, increased productivity is not the only benefit of reuse, since it becomes possible to expand the irrigated area given the availability of water and, depending on climatic conditions, multiple harvests can be carried out practically all year round. It is also worth mentioning that reuse provides significant savings in chemical fertilisers, with a reduction in environmental impact due to the reduction in contamination of watercourses, as well as relieving demand and preserving water supply.

Sunflower stands out as one of the oils with the best nutritional and organoleptic quality (aroma and flavour), and the mass resulting from the extraction of the oil results in a cake with high protein content, used in the production of animal feed. The plant can be used in silage for animal feed and its cultivation can also be associated with beekeeping. Another potential use for sunflower in the country is in the production of biodiesel, due to the growing demand from the agro-industrial and commercial sectors.

In view of the above, the general objective of this work was to analyse the influence of wastewater reuse on the microbiological quality of sunflower for animal feed. The specific objectives were to monitor the climatic conditions of the irrigated crop and to check the influence of wastewater reuse on the microbiological characteristics of the irrigated crop and whether it meets acceptable health standards. The approach taken in this work has an interdisciplinary dimension, in the social, political and economic spheres. Another important factor is the prevention of diseases in the neighbourhood of the treatment plants, as well as the formation of a reuse water market in Sergipe, an innovative subject of great interest to the state.

CHAPTER 2

2. THEORETICAL BACKGROUND

2.1. - SUSTAINABILITY AND WATER RESOURCES

According to Conway (2003), the concept of sustainability emerged in the first century AD, when a Roman landowner said: "Agriculture is a science that teaches us what crops should be planted in each type of soil, and what operations should be carried out for the land to produce the highest yields in perpetuity". It's a succinct, clear and elegant definition, but the clarity it provides has not been absorbed over time, losing its meaning and giving way to other terms to relate to the subject.

With the advent of the Second World War, the planet was faced with a model of accelerated growth in some parts of the world, especially in the countries that were involved in the conflicts. This disorganised form of growth had serious negative consequences for the environment, and humanity needed to intensify discussions about the development model and the environment.

The worldwide movement to raise awareness of environmental issues began in the 1960s and intensified after the United Nations Conference on the Human Environment (ESTOCOLMO, 1972). At the same time, in the last three decades, scientific knowledge about environmental problems has deepened, and there has been a growing realisation of the socio-economic impacts caused by these problems and even the possibility of life on the planet being compromised in the future.

The concept of a new form of development, initially called Eco-Development, was defined by Sachs (1986) as "desirable social, economically viable and ecologically prudent development". Later, the term Eco-Development was replaced by the term sustainable development, gaining prominence, especially since the *Brundtland* Report (1987) and being recognised at the 2ª United Nations Conference on the Human Environment (RIO DE JANEIRO, 1992), which brought together one of the largest numbers of heads of state in recent times and marked the incorporation of the environmental issue into the list of topics on the international negotiating agenda, continuing the tradition in 2002 (JOHANNESBURG, SOUTH AFRICA) and 2012 (RIO DE JANEIRO, BRAZIL).

In 1983, the United Nations (UN) created the World Commission on Environment and Development (WCED), an independent body with the aim of evaluating the progress of environmental degradation and the effectiveness of environmental policies. Chaired by former Norwegian minister Gro Harlen Brudtland, the World Commission on Environment and Development published one of the most important documents of our time in 1987 - the report Our Common Future, also known as the Brudtland Report (CMMAD, 1988), which was responsible for the first official conceptualisations, forms and systematisation of sustainable development, defining it as: "development that meets the needs of the present without compromising the ability of future generations to meet their own needs" (UN, 2010).

Veiga (2006), in his text, quotes Celso Furtado who presents a synthetic formula for development, differentiating it from economic growth:

> "Economic growth, as we know it, has been based on preserving the privileges of the elites who satisfy their desire for modernisation; development, on the other hand, is characterised by its underlying social project. Having the resources to invest is far from being a sufficient condition for preparing a better future for the mass of the population. But when the social project prioritises the effective improvement of the living conditions of this population, growth metamorphoses into development" (VEIGA *apud* FURTADO, 2006, p. 167).

According to Ennes (2008), post-industrial society has bifurcated the proposed solutions to the environmental issue.

> "In the context of capitalism, the issue has been dealt with through the emergence of demands generated by a concept of economic sustainability. In this sense, the environmental issue and its solutions have been incorporated into the logic of the market" (ENNES, 2008, p. 197).

Veiga (2006) states that Ignacy Sachs is one of the authors who has devoted most attention to the subject over the last few decades, since the beginning of the international controversy over the distinction between development and growth. Sachs (2008) considers that it remains valid, recommending specific objectives for eight of the dimensions: social, cultural, ecological, environmental, territorial, economic, national political and international political.

According to Sachs (2008), it is important to bring ethics and economics closer together, taking politics into account. He himself points out that the natural way to define inclusive development is by opposing the perverse growth pattern known as "exclusionary" (of the market and consumption) and "concentrating" (of income and wealth). Sachs (2008) states that education is essential for development due to its intrinsic value, as it contributes to cultural awakening, awareness, understanding of human rights, increasing adaptability and a sense of autonomy, as well as self-confidence and self-esteem.

According to the United States Geological Survey (2011), of all the water on the planet, 97% is salty and only 3% is fresh water, located in rivers, lakes, solidified in polar ice caps (68.7%), underground aquifers or in the atmosphere itself. It can therefore be seen that the availability of water resources in liquid and fresh form, which are of total importance for the survival of man on earth, is relatively low, coupled with the fact that there is an uneven distribution of fresh water around the world.

Many countries don't have enough water to meet demand and, as a result, the depletion of aquifers due to over-extraction is common. In addition, water scarcity is accompanied by a deterioration in its quality due to pollution and environmental degradation (UNEP, 2004).

Camargo (2003) states that:

> We currently live with environmental problems of different characteristics and magnitudes, such as: water pollution, atmospheric pollution, forest degradation, damage to the ozone layer, global warming, soil erosion, desertification, deterioration of species' habitats, loss of biodiversity, accumulation of toxic waste, among other problems (CAMARGO, 2003, p. 30, emphasis added).

Furthermore, according to estimates by the United Nations (UN, 2010), around 2.8 billion people will live in regions of chronic drought over the next 25 years. The UN describes water as "the oil of the 21st century" (CAMARGO, 2003, p.32). According to Brown (2003), the world is heading for a generalised water deficit and irrigation is a major contributor to this reality, due to the growth and technological evolution of water collection methods in recent decades. As a result, Câmara and Santos (2002) corroborate that in addition to irrigation being the human activity that consumes the most water, they project a figure of around 80 per cent for total world demand.

According to UNEP (2004), the outlook for the coming years with regard to water and food is unfavourable and not at all optimistic. By 2050, according to Brown (2003), countries with water deficits will experience continued population growth, condemning hundreds of millions of people to hydrological poverty. According to the UN (2012), although 89 per cent of the world's population uses treated water sources, 783 million people still lack access to drinking water, with dramatic variations by region. Only 61 per cent of people in sub-Saharan Africa have access to treated water supplies, compared to 90 per cent or more in Latin America and the Caribbean, North Africa and much of Asia.

According to the diagnosis of the Atlas Brasil - Urban Water Supply by the National Water Agency - ANA (2011), Brazil, which has the greatest water potential on the planet, is at risk of having water supply problems in more than half of its municipalities by 2015. Considering the availability of water and the infrastructure conditions of the production and distribution systems, the data reveals that in 2015, 55% of Brazilian municipalities could have a water supply deficit, including large cities such as São Paulo, Rio de Janeiro, Salvador, Belo Horizonte, Porto Alegre and the Federal District. This represents 71 per cent of the country's urban population of 125 million people, taking into account the demographic increase.

As more than 90 per cent of Brazilian households currently have access to a water supply network, scarcity seems a distant threat, as if there could be no problems in the future. This leads to the existence of a culture of water abundance, which is not true, since the internal problems of water scarcity stem from the poor natural allocation of this resource and the spatial distribution of the population, which is concentrated in certain areas (HIRATA, 2000; DNAEE, 1992).

According to the survey, the North and Northeast regions have relatively the biggest problems in water production systems. In the Northeast, the percentage of the country's water potential is 18 per cent and the region also has the greatest problems with the availability of water sources, due to the lack of rainfall (ANA, 2011).

According to Mastny and Cincotta (2005), the situation of already scarce water resources can be further aggravated or depleted, which combined with overpopulation conditions, which leads to an increase in water consumption in all scenarios, unhealthy conditions can cause deadly epidemics. In addition, the economic growth of food production becomes limited, since the poor quality of water makes it unsuitable for human, industrial and agricultural consumption (BROWN, 2003; WOLF et al., 2005).

Poor water quality is directly linked to disposal. With the increase in water consumption comes liquid waste concentrated or diluted in water (METCALF & EDDY, 2003), which must necessarily be collected and processed (or treated) in treatment systems. Worldwide, more than 80 per cent of wastewater is not collected or treated (UN, 2012). Notably, water, as a strategic element, is a fundamental part of the processes for disposing of waste generated by human activity, and it is of great importance to have advance knowledge of the types and magnitude of the damage that dumping polluting loads can cause (EIGER, 2003).

In developing countries, including Brazil, the pollution of rivers and streams by organic compounds is mainly caused by the discharge of sanitary sewage (CÂMARA and SANTOS, 2002). Therefore, there has been growing concern worldwide about the issue of scarcity related to pollution (UNEP 2004), where several countries are working to develop stricter legislation on the quality of water intended for human consumption and environmental protection (VAZQUEZ-MONTIEL et al., 1996).

According to Lucas Filho et al. (2001), the application of sanitary sewage to the soil is the simplest and one of the most efficient methods of final disposal and treatment of liquid effluents through natural processes. However, even with its great potential and list of advantages, this process has been little used in the country, although Sperling (2005) observes a growing trend towards the use of this important alternative in Brazil.

2.2. - WATER REUSE

2.2.1 - Contextualisation

According to Hespanhol (2003), once polluted water can be recovered and reused for a variety of beneficial purposes as long as it is used for less restrictive purposes. The quality of the water used and the purpose of the reuse are the factors that will define the recommended levels of treatment, the safety criteria to be adopted, the capital, operational and maintenance costs. Reuse alternatives are based on local characteristics, conditions and factors, such as political decisions, institutional schemes and availability.

The author also points out that in Brazil the most significant forms of reuse are in the urban, industrial, artificial recharge of aquifers and agricultural areas, the latter of which currently depends on water supplies at such a level that the sustainability of food production cannot be maintained without the development of new sources of supply and the proper management of conventional water resources.

The practice of distributing human and animal waste in the soil as fertiliser is ancient in South Asian countries, especially China. In the city of Athens, sewage was used for irrigation in ancient times and there are records that reuse was used in Germany and Scotland in the Middle Ages (SHUVAL et al., 1986).

According to Paganini (2003), English initiatives in the 1850s influenced the correct technique for the controlled use of sewage for agricultural purposes, with the aim of cleaning up the River Thames by separating rainwater into rivers and sewage into land farms. Nowadays, the application of effluent to the soil is seen as an effective way of controlling pollution and a viable alternative for increasing water availability, especially in arid and semi-arid regions, pointing to greater economic, environmental and public health benefits (PAGANINI, 2003).

The first reference standards developed for water reuse were drawn up by the California State Department of Public Health in 1918. Since then, the state has revised its standards, adding other possible types of reuse and the necessary treatments (CROOK, 1998). These standards were copied by many countries in arid zones around the world that needed additional water to increase agricultural production. However, since the very restrictive standards began to require the construction of very expensive and technologically advanced treatment plants, few countries could adhere to the established standards in practice.

There is currently no rigid model that should be implemented anywhere in the world, either with regard to institutional or legal issues, because international experiences are similar in some respects and different in others (RODRIGUES, 2005). In Brazil, there are references to legislation and standards on the use of effluents for reuse in agriculture. The World Health Organisation (WHO) recommends criteria for the use of wastewater. In 1973, the WHO published its first health guidelines on the use of wastewater, which were updated in 1989 and 2006.

In the latest edition of the WHO (WHO, 2006), procedures have been adopted that go beyond establishing guidelines for wastewater treatment with a view to reuse in agricultural and aquaculture activities. Less stringent standards for treatment are established, along with recommendations for handling reuse water. The recommendations are based on "health targets" established by analysing the routes of contamination (direct contact, consumption and the presence of vectors) and, on the basis of which, combined protection measures (multi-barrier protection) are recommended. In addition, parameters are established for quantitative analysis and risk measures for different exposure routes.

According to Moruzzi (2008), the removal of pathogens depends on actions related to both the treatment stage and those related to hygiene measures in the handling and application of irrigation techniques. The combination of these pathogen removal mechanisms (multi-barrier processes) can lead to the removal of up to 7 logarithmic units of indicator organisms. There are also differences relating to the type of irrigation: restricted and unrestricted. For vegetables consumed raw, crops not commercially processed, crops irrigated superficially or by sprinkling, the combination of multi-barrier processes should lead to greater degrees of protection.

Clearly, multi-barrier practices can be considered an advance in reuse water management, especially in developing countries where investment capacity is limited. However, it is worth mentioning that there is an additional difficulty related to monitoring these diffuse practices at the points of use of reuse water (MORUZZI, 2008).

However, we must beware of the risk of misinterpreting treatment as unimportant. The new guidelines should serve as a starting point in developing countries where low levels of sanitation are a reality, in the hope that advances in treatment coverage will allow for a gradual increase in treatment techniques and levels so that reuse practices can become safer (MORUZZI apud ASANO, 2008).

2.2.2 - Aspects of reuse

According to Leeuwen (1995), technologies and practices applied to minimise sewage pollution result in high quality effluents, which should be recovered rather than discarded, and seen as a possible alternative source of water to meet the demand for specific uses.

Water reuse transforms an undesirable and even harmful by-product of human activity into a useful product, even providing other benefits such as nutrients for plants when treated effluent is used for irrigation. However, the use of treated effluents in soils must be constantly monitored so that there is no contamination of the soil-water-plant system (PAGANINI, 2003).

In 1973, the WHO published a document classifying the types of reuse into different modalities, according to their uses and purposes. Indirect reuse occurs when water that has already been used one or more times for domestic or industrial use is discharged into surface water or groundwater and used again downstream in a diluted form. This is the most widespread form where the self-depuration of the body of water is used, often uncontrolled, to degrade the pollutants discharged with the sewage in natura.

Direct reuse is the planned and deliberate use of treated sewage for certain purposes such as irrigation, industrial use, aquifer recharge and drinking water. It requires the design and implementation of appropriate treatment technologies to adapt the quality of the effluent at the plant to the quality defined by the required use, while internal recycling is the reuse of water within industrial facilities, with the aim of saving water and controlling pollution. It consists of a closed-loop system where the replacement of water from another source is due to losses and the consumption of water to maintain treatment processes and operations (WHO, 1973).

According to this organisation, direct potable reuse is classified as sewage that has been recovered through advanced treatment and is directly reused in the drinking water system. It is practically unfeasible in Brazil due to the low cost of water, the high cost of treatment and the associated high health risk, while indirect potable reuse is sewage, after treatment, which is disposed of in surface or groundwater collections for dilution, natural purification and subsequent collection, treatment and finally use as drinking water. It comprises the flowchart where sewage treatment is used to bring the quality of the effluent at the plant into line with emission standards and discharge into *water* bodies.

Considering the direct reuse planned for non-potable purposes, it can be subdivided into the following modalities: non-potable reuse for agricultural purposes, which although when this type of reuse is practised there is, as a rule, recharge of the underground water table as a by-product, the purpose of the practice is the irrigation of food plants, such as fruit trees, cereals, etc? and non-food plants such as pastures and fodder, as well as being applicable for animal watering; non-potable reuse for industrial purposes, which includes the industrial uses of cooling, process water, for use in boilers, cleaning, etc. Some commercial uses can also be considered, such as washing vehicles (WHO, 1973).

The WHO (1973) classifies non-potable use for recreational purposes as being reserved for irrigating ornamental plants, sports fields, parks, lawns and also for filling ornamental and recreational ponds, etc. In urban areas, irrigation of public parks, landscaped areas, trees and shrubs along roads, fountains and water mirrors can also be considered, while non-potable reuse for domestic purposes is considered when water is reused for watering residential gardens, flushing toilets and using this type of water in large buildings. Reuse for fire reserves, washing cars and floors can also be considered.

According to the WHO (1973), reuse for flow maintenance promotes the planned use of treated effluents, with the aim of adequately diluting any pollutant loads carried by them, including diffuse sources, as well as providing a minimum flow during droughts. Reuse to maintain natural habitats can be included in this modality. Reuse in aquaculture or aquaculture consists of the production of fish and aquatic plants in order to obtain food and/or energy, using the nutrients present in the treated effluents. Reuse for recharging underground aquifers is referred to as the recharging of these aquifers with treated effluents, and can take place directly through injection under pressure, or indirectly using surface waters that have received discharges of treated effluents upstream. Recharge aims to increase the availability and storage of water, as well as to control the salinisation of coastal aquifers and to control soil subsidence.

Therefore, once properly planned, the reuse of wastewater has a positive influence on the problems of surface water pollution, which not only grants the conservation of natural sources, but also contributes to the good development of cultivated plants, since the nutrients present become available and beneficial. The nutrients present, especially phosphorus and nitrogen, reduce or eradicate the need to add commercial fertilisers, contributing to the development of agricultural production in regions with little or no water availability (PESCOD, 1992).

In addition to the nutrients and micronutrients that are absent in most of the lower-cost chemical fertilisers available on the market, the use of treated wastewater results in an increase in organic matter, which acts as a soil conditioner and increases its water retention capacity (MARQUES et al., 2003). In addition to these advantages, there is also the prevention of underground resources, soil conservation through the accumulation of humus and nutrients, and increased resistance to erosion.

According to Hespanhol (2003), studies carried out in various countries have shown that agricultural productivity increases substantially in systems that adopt irrigation with properly managed treated sewage and that, with favourable climatic conditions, it is possible to result in multiple harvests practically throughout the year (HESPANHOL, 2003a apud BARTONE and ARLOSOROFF, 1987).

As already mentioned, reuse makes it possible to minimise sewage discharges into bodies of water, resulting in benefits for public health and, consequently, a reduction in waterborne diseases caused by pathogens and chemical substances present in effluents. By integrating agricultural reuse into the collection and treatment system, effluent transport and disposal can be optimised (PESCOD, 1992). Another factor of great importance is that reuse contributes, especially in poor areas, to increased food production, raising the levels of health, quality of life and social conditions of the population.

According to Siebe (1996), the risks to human health, such as parasitic infections and/or the accumulation of heavy metals in the body transferred through the food chain, represent the greatest limitation of reuse in agriculture. However, the FAO (2003) points out that these risks are almost entirely the result of insufficient treatment of wastewater, which exposes the health of irrigation workers and food consumers, coupled with the fact that, when practised inappropriately, the agricultural reuse of sewage can cause serious problems, such as the accumulation of salts, a decrease in the infiltration capacity of the water, the accumulation of phosphate or the leaching of nitrates (PAGANINI, 2003; MARQUES et al., 2003).

According to Mota (2000), another negative effect that can occur is the accumulation of chemical contaminants in the soil. Depending on the characteristics of the sewage, the practice of irrigation for long periods can lead to the accumulation of toxic compounds, both organic and inorganic, and a significant increase in salinity in unsaturated layers. To avoid this possibility, irrigation should be carried out with sewage of predominantly domestic origin.

However, Hespanhol (2003) emphasised that the presence of pathogenic organisms in wastewater, soil or crops does not mean the transmission of diseases.

> "the protective barriers provided by factors characteristic of microorganisms (effective dose, persistence, residual load, latency, etc.), hosts (natural or acquired immunity, age and sex, general health conditions) and other factors, which mean that the actual risk of causing disease is generally much lower than the potential risk, characterised by the mere presence of pathogenic organisms" (HESPANHOL, 2003).

According to Cardoso (2005), it is important to emphasise that the existence of uncertainties regarding the rate of exposure does not negate the importance of establishing minimum criteria for the prevention of infectious diseases. Shuval et al. (1986) and WHO (1989) recommend microbiological quality criteria for the use of wastewater in irrigation.

According to Hespanhol (2003), lagoons in series, made up of anaerobic, facultative and maturation units, with average hydraulic detention times of 10 to 30 days can be designed to meet WHO guidelines (combined with the fact that Brazil has a predominantly hot climate) for both faecal coliforms and helminths, and stabilisation lagoons are the only ones recognised by the WHO for their efficiency in removing these.

According to Sperling (2005), a facultative lagoon is a simple stabilisation lagoon system that is easy to operate and low cost. It has the disadvantage of requiring large areas for construction, but in many locations in Brazil this would not be an obstacle. In the process, part of the organic matter in suspension undergoes sedimentation, constituting the bottom sludge, which undergoes the process of decomposition by anaerobic microorganisms, being converted into carbon dioxide, methane and others, with the inert fraction remaining in the bottom layer without altering its nature.

Dissolved organic matter and small suspended organic matter are broken down by facultative bacteria, which are able to survive both in the presence and absence of free oxygen. Bacteria use organic matter as a source of energy, achieved through respiration. Aerobic respiration requires the presence of oxygen, which is supplied to the environment by the photosynthesis carried out by algae. This creates a perfect balance between the consumption and production of oxygen and carbon dioxide (SPERLING, 2005).

According to Sperling (2002), with a few adaptations to the flow chart and geometric shape of the lagoons, a high degree of efficiency can be achieved in removing pathogenic organisms or, more specifically, their main indicators (coliforms and helminth eggs), even achieving significant nitrogen and phosphorus removal.

The classification of facultative lagoons, according to Sperling (2002), is as follows: primary lagoon when they receive raw sewage; secondary lagoon when they receive their effluent from a previous treatment unit, such as an anaerobic lagoon. Facultative lagoons must have a preliminary treatment system, consisting of grids to retain coarse material, a sandbox to remove inert material and a flow meter.

According to Mancuso (2003), maturation ponds are tanks that receive effluent from conventional treatment plants or other ponds and are used to improve the effluent, with the aim of reducing settleable solids and pathogenic organisms. They can be considered tertiary treatment devices, not intended to stabilise organic matter, but to improve the quality of effluent from secondary treatment facilities.

According to Daltro Filho (1997), stabilisation ponds in the state of Sergipe first appeared in the 1970s with the implementation of the Aracaju Industrial District (DIA) pond. The state of Sergipe has seventeen stabilisation pond systems in place, whose situation varies from fully operational to abandoned, deactivated or non-existent. Of these 17 lagoon systems in Sergipe, eight (08) treat domestic sewage, most of which are located in the interior of the state and in the greater Aracaju area, which includes the Roza Elze WWTP, the object of this study.

2.2.3 - Legislation

The WHO guidelines (1989) have been adopted for the agricultural use of wastewater in various countries, either in their entirety or adapted to the epidemiological, socio-economic and health policy conditions of the locality. Some countries, including developing countries, have adopted much stricter guidelines, based on the Californian standards in 1968, emphasising that their characteristic requirement is 2.2 coliforms/100 ml for unrestricted irrigation. Meeting the California requirement can only be achieved through rigorous tertiary sewage treatment processes, including disinfection.

According to Strauss et al. (2000), developing countries that have adopted these Californian standards have hardly achieved the expected results due to the fact that it is economically unviable and institutionally impossible to implement in these countries. As a result, they reuse sewage in an uncontrolled manner or definitively veto the practice due to the lack of a monitoring and control structure for its implementation (BASTOS & MARA, 1993).

The introduction of the WHO guidelines (1989) sparked a controversy over the reuse of wastewater. Questioning came from developed countries in particular, revolving around the fact that the guidelines were soft and would not achieve the objective of protecting health. Much of the opinion seemed to be based on a concept of "zero risk" (SHUVAL, 1986), resulting from guidelines or standards aimed at eliminating pathogenic organisms from wastewater. However, the WHO guidelines are based on the objective that there should be no excess infection in the population attributable to sewage reuse and that, in a specific population, it should be assessed in relation to the risks of intestinal infection from other transmission routes (STRAUSS et al., 2000).

Current recommendations by the WHO (2006) emphasise the importance of the biological quality of effluents used for irrigation, in order to reduce the likelihood of pathogens spreading and causing various diseases. Based on its studies, the WHO makes recommendations for the microbiological quality of treated sewage to be used for irrigation. In order to meet the recommendations, various forms of sewage treatment are currently used, so it was observed that stabilisation ponds have a high degree of efficiency, recognised for their excellent removal of faecal microorganisms, among other parameters, and that it is the only natural system recognised by the WHO (WHO, 2006).

In Brazil, the enactment of Law Nº 9.433/97, which instituted the National Water Resources Policy, created the National Water Resources Management System (BRASIL, 1997), which presents a new approach to the water issue, the management of water use by river basins and the concept of the paying user, which results in the rationalisation of water use, establishing principles and instruments for its use. In turn, CONAMA Resolution Nº 357/2005 defines water quality guidelines to be observed in accordance with the predominant use of watercourses, but does not deal with the whole issue of reuse.

Bill Nº 5296/2005, which establishes guidelines for the National Basic Sanitation Policy and public basic sanitation services, refers directly to water reuse. As in its article 8, item III:

> "The guidelines for sanitary sewage are: to encourage the reuse of water, the recycling of other sewage constituents and energy efficiency, subject to compliance with public health and environmental protection requirements" (BRASIL, 2005).

In 2003, the National Water Resources Council (CNRH) launched a draft resolution that was very similar to the WHO recommendation, with the aim of encouraging the reuse of wastewater and establishing effluent quality standards for the various types of reuse, representing a major step forward in legalising the technique in Brazil. However, it was not put into effect and it was only on 28 November 2005 that the agency launched Resolution Nº 54, a complementary regulation that refers to the quality standards and codes of practice for various types of reuse: reuse for agricultural and forestry purposes; reuse for urban purposes; reuse for environmental purposes; reuse for industrial purposes; and reuse in aquaculture.The regulation of water reuse is in full swing in Brazil, due to the recognition of reuse practices in the country. It should be noted that CNRH Resolution Nº 54/2005 makes water reuse an integral part of the country's current water resource management policies, but does not establish specific parameters for its use.

2.3. IRRIGATION

According to Bernardo et al. (2009), irrigation is the supply of water in such a way as to effectively meet the water needs of crops, enabling optimised development and promoting the transport of essential nutrients. In the semi-arid environment, irrigation is a sine qua non condition for agricultural production outside the rainy season. According to Mantovani et al. (2009), irrigation provides intensification in agricultural production, distributing it throughout the year, optimising the use of areas, creating jobs and promoting an increase in local income, considerably improving the population's quality of life.

According to Hespanhol (2008), irrigation consumption varies greatly, depending on the method used and factors such as the nature of the soil, the type of crop, local evaporation rates and the soil's water retention capacity, but it has high rates in relation to other uses of drinking water, which are about to reach 80% worldwide. Against this backdrop, it is essential to use this natural resource rationally, by increasing the efficiency of the techniques used or employing alternative sources of water (DUARTE, 2006).

The rational management of any irrigation system must take into account the social and ecological aspects of the region, increasing productivity and efficiency in the use of water, reducing costs, maintaining soil moisture and phytosanitary conditions favourable to the proper development of the irrigated crop. It must also improve or, at the very least, maintain the physical, chemical and biological conditions of the soil, as this will affect the useful life of the system (BERNARDO, 2009).

According to Feigin et al. (1991), irrigation methods can be classified into two broad groups: surface or gravity irrigation and pressurised irrigation. In principle, any method can be used in irrigation with wastewater, as long as the particularities in question are observed. The choice of method must follow a series of criteria, including economic considerations, topography, the physical characteristics of the soil, the types of agricultural crops to be used, the availability of labour, the quality of the water and the farming tradition of the properties.

According to Bastos et al. (2003), the physical-chemical characteristics of the effluent and the choice of irrigation method are also closely linked. High levels of bicarbonates, sodium and chlorides can limit the use of sprinkler irrigation, for example. The irrigation method expresses the rate of application of wastewater, which can lead to a faster accumulation of salts in the soil, for example in the case of flood irrigation (MENDONÇA et al., 2003).

According to WHO (2006), most of the water applied to the plant from rain and irrigation is consumed by the evapotranspiration process. Therefore, the water needed by the crop is equivalent to the amount of water spent on this process, and the rate of evapotranspiration depends on the type of crop and climatic factors, which can be estimated according to meteorological data for the region in question.

According to Allen et al. (1998), evapotranspiration is a term used to express the simultaneous occurrence of the processes of soil water evaporation and plant transpiration, controlled by the energy balance, atmospheric demand and the supply of soil water to plants. The rate of evapotranspiration from the reference surface, which occurs without water restrictions, is known as reference evapotranspiration, or ET0. The reference surface corresponds to a hypothetical pasture crop with specific characteristics. The same author emphasises that the concept of reference evapotranspiration (ET0) is used to study the demand for evapotranspiration from the atmosphere, regardless of the type and development of the crop, as well as management practices.

Also according to Allen et al. (1998), the only factors that affect ET0 are climatic parameters and it can therefore be calculated from meteorological data. To estimate ET0, the FAO recommends the Penman-Monteith method as the only one to be used based on climatic parameters.

Evapotranspiration is normally calculated from meteorological data due to the difficulties in obtaining accurate field measurements. At this point, great care must be taken with the calculation method to be used, as there are several in which some are only valid for specific climatic and agronomic conditions and cannot be applied under conditions other than those in which they were originally studied. In 1990, the FAO consulted various experts and found that the method

FAO Penman-Monteith is recommended as the standard method for defining and calculating reference evapotranspiration (ET0). This model was adopted because it provides relatively accurate and consistent results in both arid and humid climates (ALLEN et al., 1998).

The most common method for determining crop water requirements is based on estimating crop evapotranspiration (ETc), which is a two-stage process. In the first, the evapotranspiration of a reference crop (ET0) is estimated. In the second, ETc is obtained by multiplying ET0 by a crop coefficient (kc) that integrates the characteristics of the crop and the local climate (DOORENBOS and PRUITT, 1977).

Also according to the authors, they generally divide the crop development cycle for calculating crop coefficients into four phases: the initial phase, which comprises the part of the cycle from germination to when the vegetation cover reaches 10% of the surface; the second phase, which extends from 10% cover to around 70 to 80% cover; the third phase, which extends from the end of the second phase to the start of ripening (leaf fall and depigmentation); and the final phase, which extends from the start of ripening to harvest.

The crop coefficient (kc) takes on low values in the emergence phase and maximum values during the vegetative development period, which decline in the ripening phase. Pruitt et al. (1972) found that the coefficients for a plant grown under different climatic conditions and planting times can vary, as local parameters, such as temperature, relative humidity, wind and solar radiation, and the physiological and aerodynamic variations of the crop directly influence evapotranspiration.

According to Allen et al. (1998), the FAO Penman-Monteith equation is a clear, precise and simple representation of the physical and physiological factors that govern the evapotranspiration process. Climatic data on solar radiation, air temperature, humidity and wind speed are used. In order to be validated, this data has to be measured at a height of 2 metres over a large area of green pasture, with complete coverage and the soil must not be water-limited.

The process of evapotranspiration is obtained by the amount of energy available for the evaporation of water. Solar radiation is the most important energy source on the planet and has the power to convert large volumes of liquid water into vapour. Solar radiation is different for each latitude and for each season. However, it should be borne in mind that not all the available energy is used for water evaporation, as it is also used to heat the soil and the entire atmosphere (ALLEN et al., 1998). According to the same author, air temperature results from the absorption of solar radiation in the atmosphere and the heat emitted by the earth, which receives high-frequency solar radiation and emits low-frequency radiation into the atmosphere.

Allen et al. (1998) point out that the concept of relative humidity reflects the atmosphere's capacity to absorb vapour, i.e. how much water vapour the atmosphere can still take in. An atmosphere with 100% relative humidity is considered to be totally saturated and therefore unable to receive any more vapour. The higher the humidity, the lower its capacity to absorb water vapour.

> As water evaporates, the air above the evaporating surface is gradually saturated with vapour. If this air is not continuously replaced by drier air, the capacity to remove water vapour is reduced and the rate of evapotranspiration decreases (ALLEN et al., 1998).

According to Mantovani et al. (2009), determining the basic infiltration rate is an important aspect and a benchmark for sprinkler and surface irrigation. Infiltration is a process by which water penetrates the soil profile through its surface, and understanding this process is of paramount importance for the proper sizing and management of an irrigation system (BERNARDO et al., 2009).Mantovani et al. (2009) point out that infiltration of water into the soil is the process by which water penetrates its profile, usually expressed in litres or centimetres. One term that has to be taken into account in irrigation is the infiltration rate of water into the soil, expressed in cm/h or mm/h. This parameter will indicate the behaviour of a sheet of water on the soil in relation to the time it takes to infiltrate it. The infiltration process is made up of several parts: initially the infiltration rate is a function of soil moisture, soil cover and other factors; as the process progresses, it becomes a function of soil structure and texture.

2.4. SUNFLOWER CULTIVATION

According to the FAO (2007), sunflower is grown on all continents in an area of around 24 million hectares. In Brazil, sunflower cultivation has been practised in the centre-west, south, south-east and north-east states, although it is not very widespread. The crop was brought to the country by European settlers at the end of the 19th century. There have been countless attempts to encourage and expand its cultivation in different regions of the country since the beginning of the 20th century. However, in 1998, on the initiative of industries and cooperatives linked to the vegetable oil sector and since 2003, with the National Programme for the Production and Use of Biodiesel, sunflower has been included among the oilseeds intended for human consumption and vehicle energy (UNGARO et al., 2009).

The authors also point out that nowadays sunflower can also be used as a medicinal plant, a melliferous plant, a silage and fodder producer, a green manure, a soil improver and an ornamental plant. However, according to Evangelista et al. (2001), one of the problems with livestock farming in Brazil is the seasonal nature of fodder production throughout the year, leading to periods of high production followed by shortages. Therefore, in order to avoid a lack of roughage in the dry season, conservation methods have been proposed, with silage being an alternative for the country due to the periods of water deficit that hinder the production of good quality roughage and, consequently, the maintenance of animal production all year round.

The cake, a by-product of oil extraction, can also be used in animal feed, as a substitute for soya meal, and in human food, in the form of sunflower flour and "milk". Due to its high nitrogen and phosphorus content, it is an excellent source of these nutrients for soil fertilisation. Due to its importance in human and animal nutrition and as a biofuel, sunflower has been receiving special attention in terms of development mechanisms, with the aim of expanding its cultivation in the country in a rational and controlled manner (UNGARO et al., 2009).

Sunflower growth and development, from sowing to maturity (biological cycle or biocycle), is a process based on the sequence of morphological, biochemical and physiological changes that take place in the plant, conveniently considered as phenological phases, separated by phenological stages. The duration of each phase is regulated by the action of the genetic command, extrinsic to the cultivar, interacting with the environmental conditions (natural soil and climate, plus the technological level adopted) in which the crop is grown (UNGARO, 2009 apud CONNOR & HALL, 1997).

According to Leite et al. (2005), uniform sowing depth and plant distribution are key factors for growing sunflower with high yields. In order to achieve perfect germination and uniform emergence, the seed must be covered with a 3 to 5 centimetre layer of soil, the depth of the furrow varying according to the humidity and nature of the soil. Well-prepared soil allows the soil to come into close contact with the seed, allowing for adequate water absorption and guaranteeing the start of the germination process. The correct depth facilitates seedling emergence, avoiding unnecessary expenditure of energy reserves to break up the layer of soil over the seed (SEMENTES CONTIBRASIL, 1981).

According to Castro & Farias (2005), plant development is divided into two sub-periods: vegetative (V) and reproductive (R). The vegetative phase refers to the period from germination to the flower bud period and is divided into Ve (emergence) and Vi (subsequent vegetative phases), the latter characterised by the appearance of true leaves and the number of leaves at least 4 cm long (V1, V2, V3, V4 and Vn). The reproductive phase is organised as follows: R1 - the inflorescence surrounded by the immature bract becomes visible; R2 and R3 - phases related to the elongation of the internode immediately below the base of the inflorescence; R4 and R5 - phases related to the opening of the inflorescence and the start of anthesis, respectively; R6 - anthesis is complete; R7 and R8 - phases related to the filling of achenes; and R9 - physiological ripeness).

According to Ungaro et al. (2009), the sunflower's adaptation to different environments is favoured by its pivoting root system which, without physical or chemical impediment, explores deeper layers of the soil in search of water and nutrients. This allows the plant to better tolerate periods of water deficiency and to recycle nutrients.

According to Doorenbos & Pruit (1977), the most important factors in determining the crop's water requirement are: climate, crop (growth characteristic), soil moisture, agricultural and irrigation practices, and other factors that influence the growth rate (such as fertilisers), diseases and pest infestations, and invasive plants.

Ungaro et al. (2009) point out that the crop is unable to regulate its water consumption, extracting considerable amounts from the soil. When well established, it has the capacity to absorb water to a depth of 2 metres or more. Resistance to water diffusion through the stomata is low; the stomata are large, numerous and dense, mainly on the underside of the limbus.

According to Ungaro et al. (2009) apud Doorenbos & Kassam (1979), the total amount of water required by the sunflower varies between 600 and 1000mm, with fundamental dependence on the climate and cultivar, which will determine the length of the growing season and the evaporative demand of the atmosphere. According to Rawson and Constable (1980), the sunflower is a very vigorous species with the ability to obtain water, achieve very high rates of photosynthesis and maintain its activities under moderate water stress. For irrigation monitoring purposes, the effective depth of the sunflower root system, the depth at which around 80% of the amount of roots accumulated along the soil profile is concentrated, is 20 cm.

The moisture content of the soil is a limiting factor in whether or not the sunflower can extract water more easily and therefore meet its needs. As the soil dries out, it becomes increasingly difficult for plants to absorb water, due to the fact that the retention force increases while the availability of water in the soil decreases. Therefore, not all the water that the soil can store is available to plants (BERGAMASCHI, 1992).

According to Barni (1994), sunflower water requirements increase as the plant develops, from initial values of around 0.5mm/day to 0.7mm/day during the sowing to emergence phase, to a maximum of 6mm/day to 8mm/day during flowering and grain filling, with a decrease after this phase until physiological maturity. Although water consumption is low at the start of the cycle, adequate water availability during the germination and emergence phase is essential for the uniform establishment of the crop with the desired plant population.

According to the author, low temperatures lengthen the crop cycle, delaying flowering and ripening, and when they occur after flowering has begun, they can significantly affect yields. On the other hand, high temperatures during flower bud formation until the end of flowering, combined with water stress, affect pollination and fertilisation, resulting in hatched seeds, the intensity of which varies between genotypes.

Sunflower grows well between temperatures of 20°C and 25°C, with an optimum point between 27°C and 28°C, obtained under controlled conditions. However, there is no significant reduction in production in the 8 to 34°C range, which demonstrates the crop's great tolerance, withstanding regions with hot days and cold nights (UNGARO et al., 2009), According to Ungaro et al. (2009), winds above 10km/h have consequences that reduce yields, such as reduced growth and delayed plant development, smaller and fewer internodes, dwarfism of the aerial part of the plants, fewer leaves, thicker and smaller leaves, fewer stomata per leaf and their smaller size.

2.5. MICROBIOLOGICAL ASPECTS

2.5.1 - Contextualisation

According to Wanderley (2005), among the most important aspects of using wastewater for production purposes, public health is still the subject of controversy in the international technical and scientific community. Together with cultural issues, these are factors that persist in the controversy over permissible risks and, consequently, the quality of the effluent, which is necessary and sufficient to guarantee health protection.

Consensus is only evidenced by the fact that irrigation with untreated wastewater poses real risks of disease transmission and that any practice of irrigation with treated sewage involves some public health risk. However, controversy persists over the levels of risk that are admissible and, consequently, the degree of treatment and the quality of the effluent that is necessary and sufficient to guarantee health safety (HESPANHOL & PROST, 1994).

According to Bastos (1999), wastewater can contain a wide variety of pathogens. However, the mere presence of a pathogenic microorganism does not necessarily imply the transmission of diseases, but only a potential risk. The actual risk of an individual becoming infected depends on a combination of a number of factors: the resistance of microorganisms to sewage treatment, the environmental conditions of the locality, the amount of infectious dose and the pathogenicity of the infectious agents, the susceptibility and degree of immunity of the host and the degree of human exposure to transmission centres.

The World Health Organisation defends the application of these postulates in relation to human consumption of water: the microbiological risks of spreading diseases are generally of greater impact than the health risks posed by chemical substances (WHO, 2004). Care and measures must be taken to combat risks such as water contamination by bacteria, viruses, protozoa, organic and inorganic micro-pollutants and heavy metals linked to the practice of reuse in irrigation. It is necessary to establish scientific, political, institutional and legal bases for the development of national standards and codes of practice for reuse in agriculture in order to guarantee that microorganisms are not present in the water at levels of densities that are representative of the risks to users' health (BLUM, 2003).

Also according to Blum (2003), the means of control range from the application of effective treatment processes to the monitoring of water quality through periodic analyses. Therefore, sanitary control is an essential factor of the utmost importance when using this technique, paying attention to contamination and the risks to public health associated with pathogens that may be present in sewage water for reuse.

In order to minimise the risks, in 1989 the World Health Organisation defined and published specific limits for faecal coliforms and helminth eggs in sewage water for each type of crop to be irrigated, which are 1000 thermotolerant coliforms per 100 millilitres and one helminth egg per litre, respectively. It should be emphasised that interrupting irrigation with sewage two weeks before harvest is an effective measure in contaminating the crop, further reducing health risks (VAZ et al., 1996 apud WHO, 1989).

In Brazil, Resolution RDC nº . 12, of the National Health Surveillance Agency of the Ministry of Health (ANVISA) of 2nd January 2001, which regulates the sanitary microbiological standards for food, determines a maximum value of 500 NMP g^{-1} for the presence of coliforms at 45°C and the absence of Salmonella sp/25g, the same resolution calls coliforms at 45°C equivalent to thermotolerant coliforms.

The WHO edition on the use of wastewater, excreta, grey water and health was launched in 2006 and is currently in force. The current edition contains information on the global context of the burden of waterborne diseases in a population, and how the use of this type of water can contribute to the epidemiological burden. The document presents risk analysis, risk management strategies, including quantification of the various forms of health protection and strategies for implementing the guidelines (WHO, 2006).

The new version of the WHO guidelines (2006) focuses on the risks posed to workers by exposure to and direct contact with treated effluents used in unrestricted and restricted irrigation, and through the consumption of food crops eaten raw. The reduction of pathogens (viruses, bacteria and protozoa) to provide protection for those involved should be 6 to 7 logs, i.e. a load of $< 10^{-6}$ days lost per person per year, in the case of helminths, a parameter of < 1 helminth egg per litre of treated wastewater is adopted (WHO, 2006). According to WHO (2003), *DALY stands for "Deabbiiity adjusted life year"*, meaning a measure of poor health, corresponding to one year of healthy life lost.

Bacteria are prokaryotic organisms that most often appear in colonies. The component cells are diverse in shape, being spherical, spiral or rod-shaped, with sizes ranging from 0.3 to 20 iim (METCALF & EDDY, 2003). According to Paganini (1997), numerous types of bacteria are found in the intestinal flora of healthy human beings, which are routinely deposited in faeces. Due to the content of pathogenic bacteria found in the faeces of infected individuals, effluents contain a high content of varieties as well as a high concentration, such concentration varying according to the epidemiological pattern of the population in question, which is why bacteria are used as indicators of faecal pollution.

The existence of countless microorganisms in wastewater makes it difficult to isolate and identify pathogenic microorganisms, as most are numerous enough to be identified in small samples. For this reason, there are microorganisms that are indicators of faecal contamination, as they are easy to detect and indicate the presence of faecal material. For this purpose, a subgroup of total coliforms is used, called thermotolerant coliforms, i.e. they ferment in the presence of lactose with the production of gas and acid at an incubation temperature of 44.5° C ± 0.2° C in 24 ± 2 hours. Some authors suggest using the bacterium Escherichia coli, since not all thermotolerant coliforms are exclusively of faecal origin (PAGANINI, 1997).

According to Di Bernardo et al. (2002), E. coli bacteria are extremely important for assessing the efficiency of sewage treatment, since the organism indicates the presence of faeces. They are resistant bacteria and differ from other bacteria in that they ferment the lactose in the culture medium with the production of gas, in addition to the fact that determining the concentration of these pathogens is not laborious or expensive compared to other faecal organisms.

According to Silva et al. (2010), moulds and yeasts form a large group of microorganisms that originate mostly in the soil or air. The versatility of moulds can be highlighted, as they are able to assimilate any carbon source derived from food, while yeasts are more demanding than moulds. Moulds and yeasts are highly resistant to adverse conditions such as acidic pH and low water activity, with many moulds growing below pH 2.0 and yeasts below 1.5. The optimum temperature for most moulds is between 25 and 28° C. They do not grow at mesophilic temperatures (35-37° C) and rarely at thermotolerant temperatures (45 C).°

Salmonella is the main agent of food-borne illness in various parts of the world (WHO, 2005) and also in Brazil. According to Silva et al. (2010), Salmonella is a bacterium that occurs widely in animals and, in the environment, the main sources are water, soil, animal faeces, insects, etc. The disease is usually contracted through the consumption of contaminated food of animal origin, mainly beef, poultry, eggs and milk, and vegetables contaminated with manure can lead to transmission.

2.5.2 - Microbiological analyses

According to Franco & Landgraf (1996), microbiological analyses are used to verify the organisms present and to quantify them, which can measure the risks that a given food may pose to consumer health. Analyses are of the utmost importance if the microbiological standards and specifications for food are to be properly met.

There is a wide range of methods that can be used for the quantitative and qualitative detection of microorganisms in food. However, it is prudent to use methods that have been approved by regulatory bodies, whether they are standard methods or recommendations (FRANCO & LANDGRAF, 1996). According to Pelczar Jr. et al. (1997), the procedure is determined by the type of food and the specific purpose of the analysis.

According to Silva et al. (2010), microbiological food analysis aims to detect or enumerate live microorganisms. The tests used can be qualitative, which verify the presence or absence of the target microorganism(s) in a given quantity of sample, without quantifying it, and quantitative tests, which determine the quantity of the target microorganism(s) in the sample, usually per unit of mass or volume.

Silva et al. (2010) describes the plate count method as the general method used to count large microbial groups, such as mesophilic aerobes, thermophiles, moulds and yeasts, varying the type of medium, temperature and incubation time.

According to Jay (1995) and Swanson et al. (1992), food samples are homogenised, serially diluted in an appropriate diluent, plated with or on an appropriate agar medium and incubated, after which all visible colonies are counted. The procedure is based on the premise that each microbial cell present in a sample will form a separate, visible colony when provided with a medium that allows it to grow (culture medium). According to Silva et al. (2010), microbial cells can occur in clusters, making it impossible to establish a direct relationship between the number of colonies and the number of cells, so the number of colony-forming units (CFU) is used.

According to Silva et al. (2010), as described in Chapter 7 of the Compendium of Methods for the Microbiological Examination of Foods (MORTON, 2001), the standard plate count can be subdivided into: pour plate, spread plate or membrane filtration. The recommended medium for most tests is Standard Counting Agar (PCA), incubated at 35± 1 C/48±2h.°

Total plate count (Aerobic Plate Count), also known as Standard Plate Count, is the method most commonly used as a general indicator of bacterial populations in food. The process does not differentiate between types of bacteria, but it is used to obtain general information about product quality (SILVA et al., 2010).

According to Siqueira (1995), moulds and yeasts are part of a large group of micro-organisms that come mainly from the air or soil. The same author defines moulds as filamentous, multicellular fungi and yeasts as non-filamentous fungi, normally spread by insect vectors, wind and air currents.

According to Silva et al. (2010), the optimum temperature for the development of various fungi is between 25 and 28° C, with undesirable growth at temperatures between 35 and 37° C (mesophilic temperatures) and poorly at temperatures of 45° C (thermotolerant). Food spoiling moulds require oxygen for growth and are considered strict aerobes, while yeasts are able to grow in the complete absence of oxygen and in different concentrations of carbon dioxide.

Infectious fungi are rarely related to food, but certain food-borne yeasts can trigger allergic reactions and some moulds can cause infections in immunosuppressed individuals. Several moulds are producers of mycotoxins, which are toxic metabolites formed during growth (SILVA et al., 2010).

According to Silva et al. (2010), the most probable number technique is a quantitative analysis methodology that allows the most probable number (MPN) of the target organism(s) in the sample to be determined by inoculating aliquots of this sample into a series of tubes containing a liquid culture medium suitable for their growth. Determining the number of microorganisms is based on the principle that by subdividing the sample into aliquots, some will contain microorganisms and others will not, depending on the quantity of microorganisms in the sample.

The number of aliquots with microorganisms and aliquots without microorganisms allows for a probabilistic estimate of the original density of microorganisms in the sample. The application of probability theory depends on the organisms being randomly and homogeneously distributed throughout the sample, which in the case of solid samples can be achieved by preparing and homogenising the first dilution and taking aliquots from this dilution. There are cases where aliquots of the solid sample are inoculated directly into the culture broth, but this is rarer and depends on the type of sample (SILVA et al., 2010).

The authors also point out that since inoculation is carried out in liquid media, the MPN technique has several advantages over standard plate counting. The first consists of the possibility of inoculating larger quantities of the sample, proportionally increasing the volume of culture medium, giving the technique greater sensitivity than plate counting and great flexibility in establishing the detection limit. Another advantage is that the technique allows the introduction of injury recovery steps, using a non-selective medium for the initial inoculation, which is more favourable to the injured microorganisms, and then transferring the culture to selective media.

Silva et al. (2010) classifies the MPN technique as versatile, allowing for the enumeration of different groups or species of microorganisms by varying the culture medium and incubation conditions; in this study it will be used to count thermotolerant coliforms and E. coli. Another application would be to adapt qualitative methods to quantitative ones, such as Salmonella counting (presence/absence technique).

Standard procedures are used to homogenise the sample and prepare the dilutions, and there are two formats for inoculation, depending on the aliquots. One is the multiple dilution test format and the other is the single dilution test format, in which all the aliquots inoculated are of the same dilution, with an equal amount of the sample (SILVA et al., 2010).

The traditional method for analysing Salmonella is considerably sensitive, limited to the detection of one colony-forming unit/25g of sample, but it is slow and labour-intensive (SILVA et al., 2010). In the last ten years there has been a great advance in the development of new methods, forming analytical "kits" with a registered trademark defined by the AOAC (Assossiation of Official Analytical Chemists) as: "systems containing all the key components for carrying out the analysis of one or more microorganisms, in one or more types of food, according to a given method" (SILVA et al., 2010 *apud* ANDREWS, 1997). The great thing about kits is that all or part of the material needed for the tests is sold together, without the need for preparation in the laboratory.

CHAPTER 3

3. RESEARCH METHODOLOGY

3.1. - CONTEXTUALISING THE RESEARCH

The experiment was conducted in a greenhouse at the Department of Agronomic Engineering (DEA), located at the Federal University of Sergipe, in São Cristóvão/Sergipe, under the geographical coordinates of 10° 55'46"S latitude and 37°06'13"O longitude, at an altitude of 8 m (Figure 3.1).

The greenhouse used in the experiment has a simple arched roof model which, according to Martinez (1997), is easy to build, low in maintenance costs, has a high coefficient of utilisation of the sun's rays and is practical when laying plastic.

The dimensions of the experiment site are 5.30 metres wide, 12.20 metres long and with a ceiling height of 3.00 metres, covered with 0.10 mm thick transparent low-density polyethylene to protect against rain and shade screens on the sides to ensure ventilation. The experiment was carried out on 03 metal benches with a height of 0.45 metres and dimensions of 2.06 x 1.25 metres.

Figure 3.1: Front view of the greenhouse used in the experiment (UFS/DEA).

Source: Roseanne S. de Carvalho (2012)

Site preparation was carried out before sowing, consisting of removing grasses and weeds from around and below the metal benches, using hay mowers to obtain a clean crop. Sowing took place on 03/07/2012. The seeds were placed in a row in soil furrows, in a number of five (05) units. The first thinning of the seedlings took place ten days after sowing, removing the least vigorous ones.

The experiment took place between July and September 2012 and sunflower plants (Helianthus annuus L.) were grown in cone-shaped plastic pots (top diameter 29.0 cm, bottom diameter 16.5 cm and height 50 cm), with a volume of 22.08 dm^3 , irrigated daily and containing seeds supplied by EMBRAPA Tabuleiros Costeiros. The treatments used differed in the proportions of treated effluent and water from the Sergipe Sanitation Company (DESO). The growing cycle lasted 70 days after sowing.

The treated effluent used in the experiment came from the Rosa Elze Wastewater Treatment Plant (WWTP), located in the Rosa Elze neighbourhood, in the municipality of

São Cristóvão, state of Sergipe. The WWTP treats the wastewater generated by the Rosa Elze and Eduardo Gomes neighbourhoods, operating at a flow rate of approximately 7.6 $L.s^{-1}$, made up of five stabilisation ponds arranged in series, two of which are facultative and three are maturation ponds, covering a total area of 29,650m^2 . The Rosa Elze WWTP was built in the 1980s and is maintained and operated by DESO. The physical characteristics are shown in Table 3.1.

Table 3.1 - Characteristics of the Rosa Elze WWTP lagoons

Lagoon	Depth (m)	Area (m)²	Volume (m)³
Primary optional	2,00	8.735	17.470
Optional secondary	1,98	6.962	13.785
Maturation 1	1,96	4.712	9.236
Maturation 2	1,94	4.618	8.959
Maturation 3	1,92	4.623	8.876

Source: Floor plan of the Rosa Elze stabilisation pond system project (supplied by DESO 2012).

The WWTP used in the study is fed by sanitary sewage at two points: one at the primary facultative lagoon, which represents the largest contribution to the system, according to information from DESO, receiving sewage from the pumping station; the other at the secondary facultative lagoon, which receives sewage by gravity. At both points, the sewage arrives at the pre-treatment unit, consisting of a grid and sandbox, and is then sent to the lagoons (Figure 3.2).

Figure 3.2: Raw sewage
Source: Roseanne S. de Carvalho (2012)

According to Mendonça et al. (2005), studies carried out at the Rosa Elze WWTP show that there are no high-risk pathogens or protozoa due to the fact that the system's detention time is high, around 141 days, favouring excellent efficiency in removing parasites that settle out during wastewater treatment. Therefore, this study focused on bacteria (thermotolerant coliforms, E. coli and Salmonella) and moulds and yeasts, due to the silage condition of the material.

After treatment, the wastewater was collected for the experiment in 20-litre plastic pots, as shown in Figures 3.3 and 3.4:

Figures 3.3 and 3.4: Collection point
Source: Roseanne S. de Carvalho (2012).

Therefore, the sources used in the experiment were twofold: drinking water from DESO, collected in a 500-litre reservoir located next to the greenhouse, and treated wastewater from the Rosa Elze WWTP, transported weekly to the experiment site in 20-litre plastic reservoirs with lids. The amount of water collected each week was in accordance with the water requirement, determined according to each of the four phenological phases of the crop, in accordance with FAO (1998).The soil for the experiment was prepared according to the crop's needs, with a view to favouring seed germination and the development of the plant's root system, and was sourced from the town of Umbaúba, in the municipality of Sergipe, in an area belonging to EMBRAPA TABULEIROS COSTEIROS.

3. 2 EXPERIMENTAL DESIGN OF IRRIGATION WATER AND IRRIGATION SYSTEM

The experimental design was entirely randomised (DIC), consisting of five treatments with four replicates. The treatments consisted of the proportions described in Table 3.2.

Table 3.2: Proportions used in sunflower irrigation treatments

Treatment	Proportions used
T1	100% DESO water
T2	100% effluent
T3	50% DESO water + 50% effluent
T4	25% DESO water + 75% effluent
T5	75% DESO water + 25% effluent

Source: Roseanne S. de Carvalho (2012)

The effluent and water were distributed over the treatments by means of a daily irrigation system and 100% of the crop's initial evapotranspirometric demand was replaced individually in each pot (surface irrigation with a watering can). The reference evapotranspirometric demand was estimated daily using the standard FAO 56 Penman-Monteith method (Equation 3.1) and the crop coefficient (Kc), which was multiplied by ETO, resulting in crop evapotranspiration (ETpc) (Equation 3.2).

$$ET_0 = \frac{0{,}408\Delta(R_n - G) + \gamma\frac{900}{T+273}U_2(e_s - e_a)}{\Delta + \gamma(1 + 0{,}34U_2)} \qquad \text{(Equação 3.1)}$$

The variables are:

ET0 = Reference evapotranspiration, mm.day^{-1}
Δ= slope of the saturation vapour pressure curve, kPa.°C^{-1}
Rn = surface radiation balance, MJ.m2.day
G = ground heat flux, MJ.m2.day
γ = psychometric constant, kPa.°C^{-1}
T = air temperature measured at a height of two metres, °C
U2 = wind speed measured at a height of two metres, m.s
es = water vapour saturation pressure, kPa
ea = current water vapour pressure, kPa

$$ETpc = Kc \cdot ETo \qquad \text{(Equação 3.2)}$$

Equation 3.2 is used when the crop is irrigated by sprinklers or surface irrigation, as they generally wet the entire cultivated area. When using drip or micro-sprinkler irrigation, which wets a fraction of the cultivated area, the equation is increased by an adjustment factor called the localisation factor (fL) (FACCIOLI, 2006).

The meteorological variables (temperature, relative humidity, solar radiation and wind speed) were obtained daily from an automatic weather station (Figures 3.5 and 3.6) installed inside the greenhouse and the crop coefficient for sunflower was defined in FAO document 56 (1998).

Figures 3.5 and 3.6: Display and automatic weather station

Source: Roseanne S. de Carvalho (2012)

3.3COLLECTING , PREPARING THE SOIL AND CONDUCTING THE EXPERIMENT

The soil used in the experiment was collected in the town of Umbaúba, located in the state of Sergipe, on EMBRAPA property, on 19/04/2012. The collection procedure basically consisted of not de-characterising the material, removing it in layers 20 cm thick (0-20cm; 20-40cm and 40-60cm) and arranging them in the same sequence in the plastic pots (Figures 3.7 and 3.8).

Figures 3.7 and 3.8: Soil arranged in the plastic pots and Covering the pots to be transported
Source: Danielle Gomes (2012).

After the pots were finally placed in the greenhouse, the soil was humidified for 48 hours, identified and collected for characterisation. At the same time, the metal benches were reinforced to support the weight of the whole assembly.

According to Souza (2004), the availability of nutrients in the soil is one of the environmental factors that limits the production and development of plants and in their absence the plant cannot complete its life cycle. They are divided into macro- and micronutrients, depending on the quantities required by the plants, which, if not met, will form limiting factors for the sunflower's vegetative growth and the plants' dry matter production, which can lead to visual symptoms characterising nutritional deficiencies.According to the results of the chemical analysis of the soil, the recommendation for planting fertiliser was made on 27/06/2012, with quantities of 0.4416 grams of potassium (K) and 1.104 grams of phosphorus (in the form of P_2O_5) per pot, made together with the soil moistening itself (planting fertiliser).

Boron (B) and zinc (Zn) were increased to the following amounts per pot: 550 mg of boric acid and 675 mg of zinc sulphate (BOLETIM 100/SP). Boron deficiency is more common than deficiency of any other micronutrient. Sunflower is one of the most sensitive crops to B deficiency and can be used as an indicator plant for the level of B availability in the soil (SCHUSTER; STEPHENSON apud SOUZA, 2004) (Figure 3.9).

Figure 3.9: Correcting the soil with boron and zinc.
Source: Danielle Gomes (2012).

On the day before sowing, preparatory irrigation was carried out, with around 1 litre of DESO drinking water, and after sowing, twice a day (500 ml) for a week. As a high amount of water was observed percolating through the pot, saturation irrigation was carried out, each consisting of 335 ml of DESO drinking water, the value was obtained by carrying out the percolation test, which consisted of placing trays below the pots to obtain the percolated volume, the aim of this type of irrigation was to promote uniform germination and good root development of the plants.

The seeds were sown on 3 July 2012. Five seeds were provided, recommended by EMBRAPA TABULEIROS COSTEIROS and bearing personal information from the said institution, which were of the hybrid type (Figure 3.10). The seeds were placed in a row in each pot, in shallow furrows, spaced about 1.5 cm apart and with an average depth of 2.0 cm from the surface.

Figure 3.10: Seeds laid out on the ground
Source: Roseanne S. de Carvalho (2012).

About three days after planting, the first seeds germinated (Figure 3.11). Ten days later, the first thinning was carried out, removing three seedlings from each pot, with preference given to seedlings with possible fungal presence and less vigour. On 1 August 2012, the final thinning of the experiment was carried out, leaving a single seedling in each pot.

Figure 3.11: Start of seed germination
Source: Danielle Gomes (2012).

After 15 and 36 days after sowing, the first and second top dressings were applied respectively, the latter consisting of a solution of 2.45 g of urea and 500 ml of DESO water per pot (Figure 3.12).

Figure 3.12: Top dressing
Source: Roseanne S. de Carvalho (2012).

At 56 days after sowing, the flower buds began to appear on the plants and, at 70 days, when more than 50% of the plants had flowered (Figures 3.13 and 3.14), all the material (20 plants) was cut off close to the ground for microbiological analysis of the sunflowers, analyses of Thermotolerant Coliforms, E. coli, Moulds and Yeasts and Salmonella (Figure 3.15). The material was collected because the crop was at the beginning of the reproductive phase. The material was duly packed in 10kg paper bags and labelled (Figure 3.16), and placed in an oven for 48 hours at a temperature of 65° C, where it remained until it reached constant weight (dehydrated). The plants were then properly ground and packed in sterile plastic bags (Figure 3.17).

Figures 3.13 and 3.14: Flower bud initiation and flowering

Source: Roseanne S. de Carvalho (2012).

Figures 3.15 and 3.16: Thinning plants for microbiological analysis and preparing plants for the greenhouse
Source: Danielle Gomes (2012).

Figure 3.17: Grinding process
Source: Danielle Gomes (2012).

3.4 PRESENCE OF PESTS AND DISEASES

Less than a month after sowing (19/07/12), the experiment saw the presence of the leaf-mining fly pest, mealybugs and fungi. The powdery mildew fungi (Figures 3.20 and 3.21) were combated with fungicides based on sulphur and mancozeb, in the proportions of 0.675 g per 750 ml of drinking water, applied with a hand sprayer only to the leaves of the seedlings in fortnightly cycles (Figure 3.22). After applying the fungicide, irrigation was carried out with a treatment of 335 ml of drinking water twice a day.

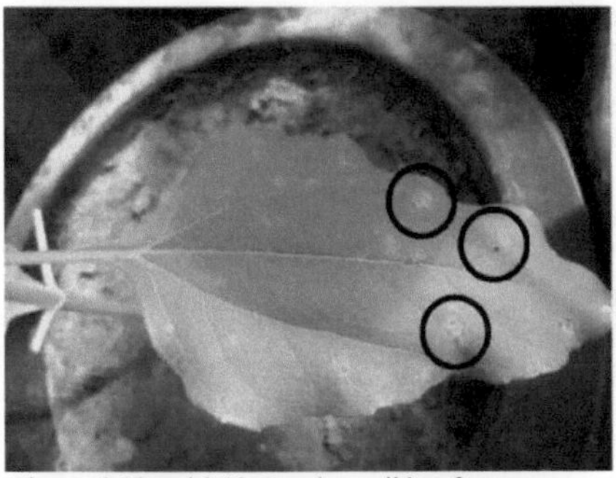

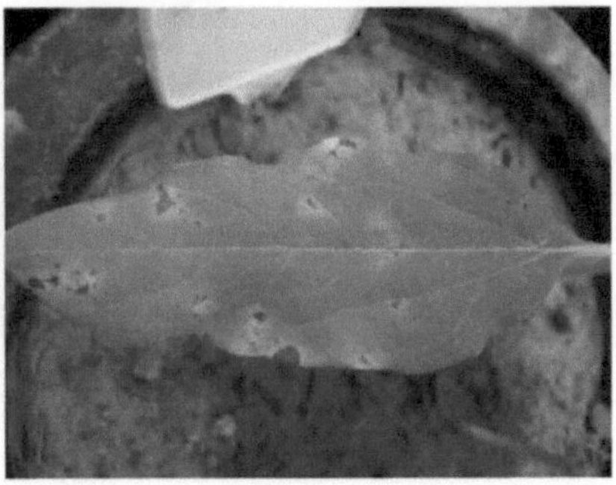

Figures 3.18 and 3.19: Powdery mildew fungus
Source: Danielle Gomes (2012)

Figure 3.20: Spraying
Source: Roseanne S. de Carvalho (2012).

After the top dressing irrigation, all the leaves showing symptoms of leaf miner fly were removed to combat proliferation. Figure 3.23 shows the larvae and eggs. Caterpillars were observed throughout the experiment (Figure 3.24) and were removed as soon as they were detected.

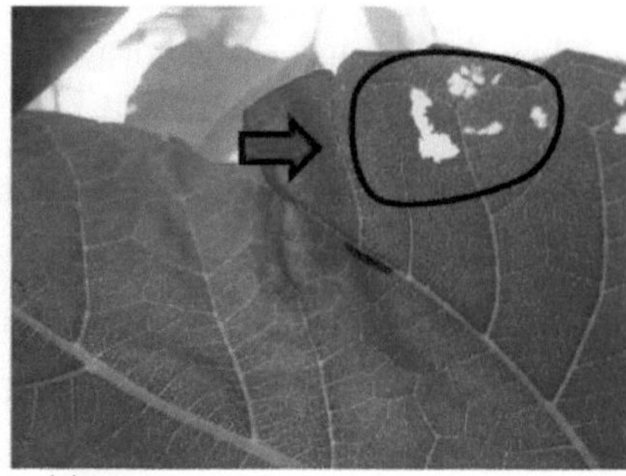

Figures 3.21 and 3.22: Worm and caterpillar eggs and larvae and damage
Source: Danielle Gomes (2012).

3.5ANALYSIS OF RAW WASTEWATER, TREATED WASTEWATER AND DRINKING WATER

The physico-chemical determinations were carried out at the Water Laboratory of the Sergipe State Institute of Technology and Research (ITPS). The total coliform and thermotolerant coliform indices of the samples collected were determined using the method described in the Standard Methods for the Examination of *Water and Wastewater, 21st* ed. Washington (2005).

The analyses of the treated and wastewater used in the experiment were carried out at the Sergipe State Research and Technology Institute (ITPS) and the results were analysed in accordance with CONAMA Resolutions° 357 of 17 March 2005 and° 430 of 13 May 2011. Two analyses were carried out on the treated water, one before the sunflowers were sown and the second at harvest. With regard to wastewater, seven analyses were carried out throughout the experiment, the first before sowing and the others distributed over the months, intensifying at the end of the experiment.

3.6 . MICROBIOLOGICAL ANALYSES OF THE CULTURE

3.6. 1Sample preparation

The preparation of samples for microbiological analyses was based on the guidelines contained in the *American Public Health Association (APHA)*, described in the 4ª edition of the *Compendium of Methods for Microbiological Examination of Foods* (DOWNES & ITO, 2001). The samples were prepared in three stages: homogenisation of the contents and removal of the analytical unit, preparation of the first dilution of the analytical unit and preparation of serial decimal dilutions for inoculation into the culture media (SILVA et al., 2010). The entire procedure, from preparation to obtaining the results, was carried out by ITPS, an organisation duly accredited by INMETRO.

In the first stage, the samples were homogenised and analytical units of 25 grams each were taken, in accordance with ISO 6887-1 (1999) and the Compendium. 04 samples were taken from each of the five treatments, making a total of 20 samples. The remaining material was stored under the same conditions as before the analyses (SILVA et al., 2010).

According to Silva et al. (2010), in the second stage of sample preparation, the analytical unit was diluted and homogenised with diluent to allow inoculation into the specific culture media for each analysis; in the experiment, 0.1% peptone water was used as the diluent, as recommended by the Compendium itself. The initial dilution was 225 ml of diluent for every 25 g of sample. ISO 6887-1 (1999) recommends that the temperature of the diluent should be the same as the ambient temperature to avoid damaging the microorganisms through thermal shock.

Serial dilutions of samples aim to reduce the number of microorganisms per unit volume, allowing for counting at the end of the process, and are generally decimalised to simplify calculations. According to ISO 6887-1 (1999), the entire procedure must not exceed 45 minutes and the interval between the end of the preparation of the first dilution and the start of the second and subsequent dilutions must not exceed 30 minutes. The process for the second and subsequent dilutions consisted of aseptically transferring 1 ml of the first dilution (10^{-1}) into 9 ml of diluent, taking care to shake 25 times in a 30 cm arc for about 7 seconds before each transfer.

3.1.1 Mould and yeast count

The analysis was carried out using the pour plate method. Samples and serial dilutions were first prepared, and for inoculation, three dilutions were selected and 0.1 ml of each dilution was inoculated into separate, sterile, empty Petri dishes.

To add the culture medium, 12 to 15 ml of potato agar, previously melted and cooled to 44-46° C, was poured onto the inoculated plates. The inoculum was mixed with the culture medium by gently moving the plates on a flat surface in figure-of-eight or circular movements eight to ten times clockwise and anti-clockwise. The plates were spread out on a bench so that the medium solidified without stacking and incubated at 35° C for 48 hours.

To count the colonies, plates with 25 to 250 colonies were selected and counted using a magnifying glass on a colony counter. The number of colony-forming units (CFU) per gram of sample was calculated by multiplying the number of colonies by the inverse of the inoculated dilution (CFU.g^{-1}) (MORTON, 2001).

3.1.2 **Thermotolerant Coliform and** Escherichia coli **count**

The American Public Health Association (APHA) most probable number method was used. After preparing the samples and serial dilutions, three dilutions of the sample were selected and inoculated in a series of three tubes of Lauryl Sulphate Tryptose Broth (LST) per dilution, adding 1 ml of the dilution per tube with 10 ml of LST. The LST tubes were incubated at 35° C for 24 to 48 hours until gas was produced. When gas was produced, a well-loaded portion of each culture was transferred to E. Coli Broth (EC) tubes and incubated for 24 hours in a water bath at 45° C until gas growth was observed (positive reading) (SILVA et al., 2010 apud KORNACKI and JOHNSON, 2001). The number of CE tubes with gas production was recorded and the Most Probable Number (MPN)/g determined according to the tables (SWANSON et al., 1992).

For the E. coli count, each EC tube with gas production within 48 hours was streaked with the culture on Levine Eosin Blue Methylene Agar (L- BEM) plates and incubated at 35° C for 24 hours and observed for the development of typical E. coli colonies, nucleated with a black centre, with or without a metallic sheen.

Two typical colonies were transferred in isolation from each plate to inclined PCA tubes and incubated at 35° C for 24 hours. Gram staining was carried out and they were inoculated into test medium and the number of tubes of EC broth that confirmed the presence of E. Coli was noted and the MPN/g determined according to the tables.

3.1.3 **AOAC Method** (Association of Official Analytical Chemists)

It was carried out in the form of a wet composition, with each analytical unit being pre-enriched separately and the pre-enrichment broths obtained being composed. The volume of selective enrichment broth (Rappaport-Vassiliadis Broth - RV or RVS) was made in the proportion 0.1ml of pre-enrichment to 10ml of RV or RVS, so 200ml of RV or RVS was required.

CHAPTER 4

4. RESULTS AND DISCUSSIONS

4.1. CLIMATE

According to Ungaro et al. (2009), various studies have been carried out to quantify the effects of the environment on the growth, development and yield of cultivars. The authors highlight three variables considered to be the most important in the environment, which are light (solar radiation), water availability and temperature, so the bioclimatic requirements of the species associated with the variables described, added to the physical characteristics, define the capacity for storing water in the soil and indicate the times and areas in which each plant species can be grown, resulting in maximised physical yield and reducing risks. It should also be pointed out that, when it comes to climate risks related to sunflower cultivation, it is important to also characterise and consider the limitations imposed by atmospheric conditions on the occurrence of the main diseases.

Throughout the experiment, it can be seen that temperatures reached a maximum of 30.53° C and a minimum of 20.99° C, with average temperatures reaching around 25° C (Table 4.1), which are therefore acceptable values for the good development of the crop in question, as shown in Figure 4.1.

Table 4.1: Monthly averages of temperature, relative humidity and radiation inside the greenhouse.

Month	Maximum T. °C	Average temperature °C	Minimum temperature °C	Humidity Relative (%)	Radiation (W/m)²
July	28,95	24,50	21,17	81,30	153,06
August	29,34	24,49	20,99	78,95	159,58
September	30,53	25,43	21,49	76,05	151,73

Source: Roseanne S. de Carvalho (2013)

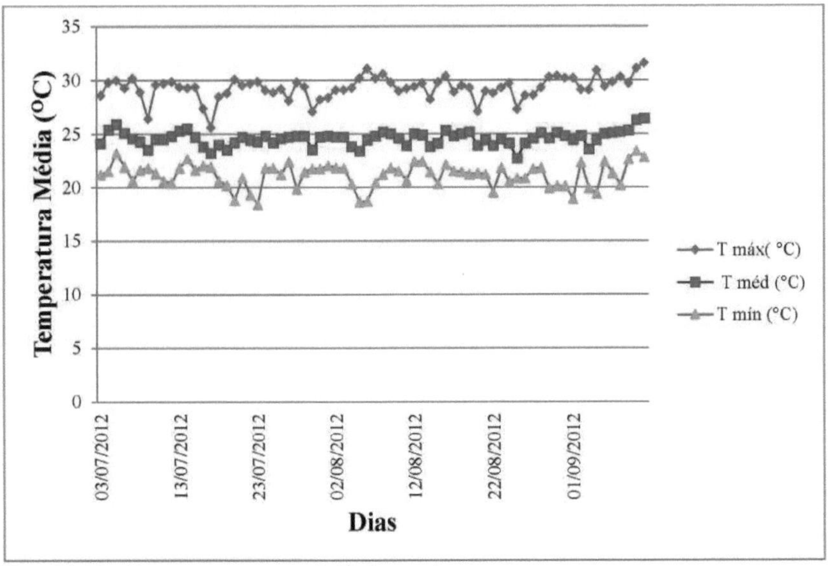

Figure 4.1: Monthly temperature averages inside the greenhouse.
Source: Roseanne S. de Carvalho (2013)

Ungaro et al. (2009) in their studies point out that the speed of germination and plant emergence increases exponentially as the temperature rises from 3 to 30° C. However, temperatures between 37 and 40° C (above 35° C) are detrimental to the germination of the sunflower crop and when the temperature reaches 45° C the seeds fail to germinate. Sunflower grows well at temperatures ranging from 20° C to 25° C, although there are studies that indicate that under controlled conditions, temperatures between 27° C and 28° C are considered optimal (UNGARO et al., 2009). Connor & Hall (1997) point out that the optimum temperature is around 26° C and the maximum is around 40° C. However, Castro and Farias (2005) note that there is no significant reduction in production in the range of 8° C to 34° C, which demonstrates the crop's high tolerance, withstanding regions with hot days and cold nights.

According to Barni (1994), low temperatures result in an increase in the crop cycle, delays in flowering and ripening and when they occur after flowering has started, they can directly affect the crop yield. Castro and Farias (2005) corroborate this by pointing out that during initial development, low temperatures can cause leaf deformities and damage the apex of the plant. The authors emphasise that temperatures below 4° C to 5° C do not cause physiological activity. Ungaro et al. (2009) point out that numerous studies have determined the effect of temperature on the sunflower cycle, so it is worth emphasising its importance for this study.

As can be seen in Table 4.1 and Figure 4.2, solar radiation during the experiment averaged 153.06, 159.58 and 151.73 W/m^2 for July, August and September, respectively, with a peak only at the beginning of August for a few days. Relative humidity (Figure 4.3) was 81.30, 78.95 and 76.05 per cent for July, August and September, respectively, in other words, showing an average decrease over the course of the experiment, although both solar radiation and humidity can be considered acceptable values for sunflower crops.

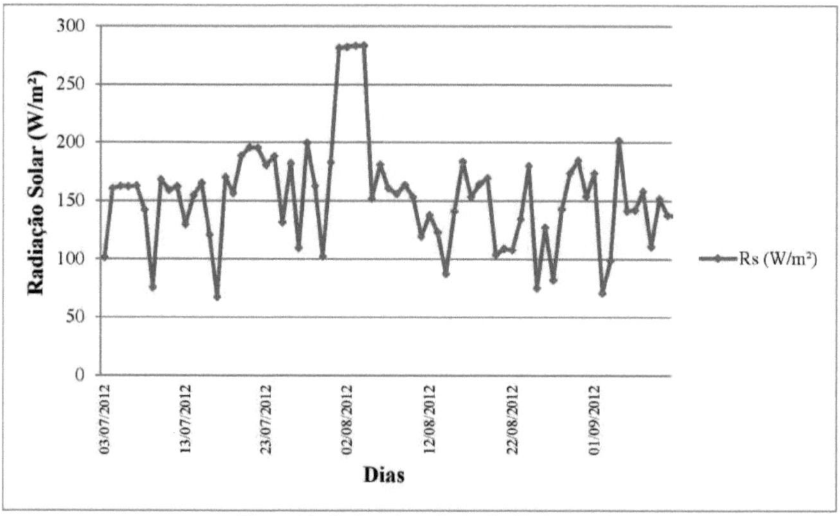

Figure 4.2 - Daily solar radiation inside the greenhouse
Source: Roseanne S. de Carvalho (2013)

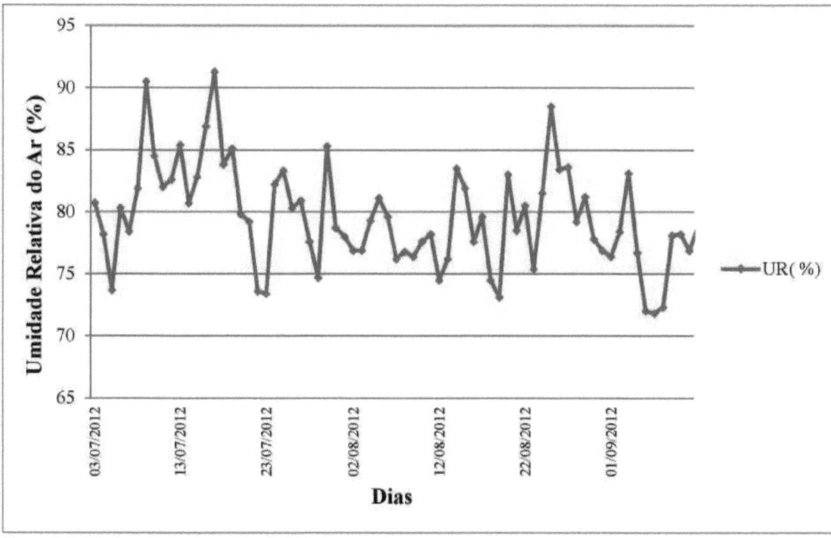

Figure 4.3 - Daily relative humidity inside the greenhouse
Source: Roseanne S. de Carvalho (2013)

Heldwein et al. (2012) show in their studies that global solar radiation (Rg) has a high influence on productivity in all crops, including sunflower, because it is directly related to the amount of carbohydrates from photosynthesis. The authors also emphasise that other processes relating to the plant and the environment depend on radiation, such as the transfer of water from the surface to the atmosphere, the cooling and heating of the soil and air and the process of evapotranspiration, the latter of which will be detailed later in this paper. Global radiation interacts on the earth's surface according to various factors, such as latitude, altitude, time of year, soil water availability and surface temperature, among others, giving rise to the radiant energy available to the vegetated surface or the radiation balance. This energy is a determining factor in estimating crop water losses through the process of evapotranspiration (MONTEITH & UNSWORTH, 1990).Monteith (1977b) in an article established the basis for comparing the empirical relationship between the accumulation of dry matter and the accumulation of solar radiation intercepted by a crop. By compiling experimental results obtained under different environmental conditions, the same author concluded that most crops accumulate around 1.4 g of dry matter per MJ of solar radiation intercepted. The article was particularly important in that it presents the efficiency of radiation utilisation as an independent and suitable parameter for describing crop performance and highlighting limitations in productivity (FERREIRA, 2007).

4.2. IRRIGATION

According to FAO 33 (2004), the sunflower crop is tolerant to water deficit and partially recovers from stress, showing less proportional reductions in yield with the use of reduced water. The variables collected from the automatic weather station were used in the Penman-Monteith equation to obtain the ET0 and ETpc values, which are shown in Figure 4.4. You can see the behaviour of the figure in relation to solar radiation and crop evapotranspiration, both of which are similar, so it can be concluded that reference evapotranspiration is closely linked to the solar radiation incident on the crop, i.e. it is a preponderant factor in obtaining ET0.

Still in relation to Figure 4.4, when relating reference evapotranspiration and crop evapotranspiration, it can be seen that in the experiment both were increasing, due to the fact that the crop was developing, since the climatic conditions per phenological phase were practically constant, so as the crop develops its leaf area, it consequently transpires in greater quantities, requiring greater replacements. The fact that the end of the experiment materialised at harvest time means that the figure does not visualise the crop's senescence phase, i.e. evapotranspiration decreasing.

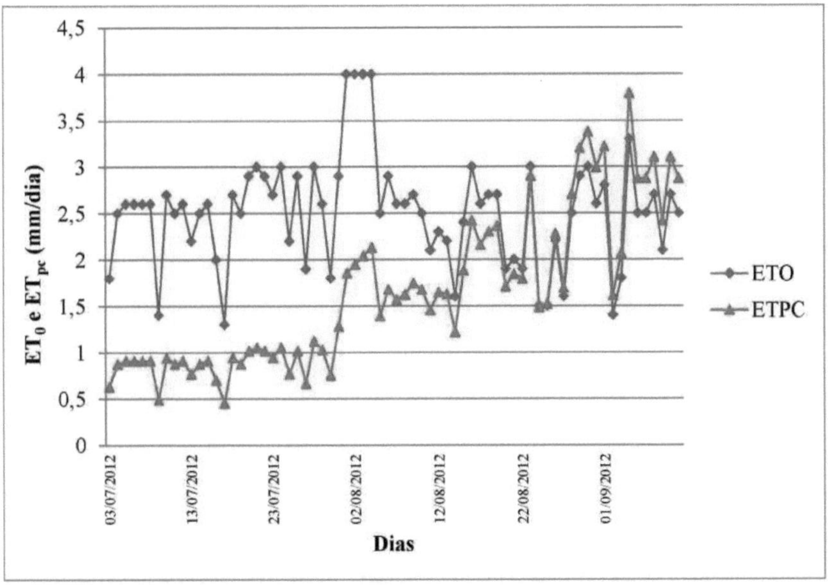

Figure 4.4 - Daily reference evapotranspiration (Et0) and crop evapotranspiration (Etpc) for sunflower.
Source: Roseanne S. de Carvalho (2013).

The total reference evapotranspiration over the 70-day cycle was 219.30 mm per pot, while the crop evapotranspiration was 165.80 mm per pot and the water demand was 7.65 litres per pot.

4.3. CHEMICAL QUALITY OF TREATED EFFLUENT AND DRINKING WATER

The chemical analyses of the treated water and wastewater used in the experiment comprised: BOD5, pH, total phosphorus and total nitrogen. The results are shown in Tables 4.2 and 4.3.

Table 4.2 - Results of treated water analyses

Sample	Date of collection	BOD (Respirometric Method)	Total Nitrogen	Total Phosphorus (RBLE)	pH
Treated water 01 (DESO)	27/06/12	91.0 mg $O2L^{-1}$	0.3942 mgL^{-1}	0.04 mg PL^{-1}	7,12
Treated water 02 (DESO)	10/09/12	97.0 mg $O2L^{-1}$	Not detected	0.02 mg PL^{-1}	7,51

Source: Adapted from ITPS (2012)

Table 4.3 - Results of the analyses of treated wastewater (RA)

Sample	Date of	BOD	Total Nitrogen	Total Phosphorus	pH

	collection	(Respirometric method)		(RBLE)	
AR 01	27/06/12	102.0 mg O2L^{-1}	21.62 mgL^{-1}	2.30 mg PL^{-1}	7,85
AR 02	19/07/12	183.0 mg O2L^{-1}	22.07 mgL^{-1}	2.52 mg PL^{-1}	7,12
AR 03	23/08/12	91.0 mg O2L^{-1}	20.16 mgL^{-1}	2.65 mg PL^{-1}	7,53
AR 04	03/09/12	173.0 mg O2L^{-1}	22.58 mgL^{-1}	2.62 mg PL^{-1}	7,73
AR 05	06/09/12	135.0 mg O2L^{-1}	23.26 mgL^{-1}	2.51 mg PL^{-1}	7,61
AR 06	10/09/12	118.0 mg O2L^{-1}	24.44 mgL^{-1}	2.56 mg PL^{-1}	7,69
AR 07	20/09/12	162.0 mg O2L^{-1}	21.29 mgL^{-1}	2.63 mg PL^{-1}	7,65
Average	---	137.7 mg O2L^{-1}	22.20 mgL^{-1}	2.54 mg PL^{-1}	---

Source: Adapted from ITPS (2012)

Biochemical Oxygen Demand (BOD5) is one of the main parameters for determining water quality. BOD measures the amount of oxygen needed to oxidise biodegradable organic matter under aerobic conditions, i.e. it assesses the amount of dissolved oxygen (DO) in mg L^{-1} of o2, which will be consumed by aerobic organisms when degrading organic matter. Therefore, BOD5 is a water quality variable that quantifies organic pollution due to oxygen depletion, which can lead to anaerobic conditions in the aquatic ecosystem.

According to CONAMA Resolution° 357/2005, rivers are classified into different classes according to their physical-chemical and microbiological characteristics, and one of the differentiating parameters is BOD5. Class 1 will therefore be used for DESO treated water and class 3 for treated wastewater as it is used for irrigating tree, cereal and fodder crops. According to the standard referred to, class 1 rivers must have BOD5 < 3 mg o2 L^{-1} , the results presented were 91 and 97 mg o2 L^{-1} , not meeting the parameters mentioned, the fact that the water was not collected directly from the tap but from a reservoir located next to the vegetation house, a reservoir that could contain some type of fouling.

With regard to treated wastewater, CONAMA Resolution° 430/2011 states that the maximum Biochemical Oxygen Demand - BOD for 5 days, 20°C can be 120 mg o2 L^{-1} , so samples AR 01, AR 03 and AR 06 with values of 102.0, 91.0 and 118.0 mg o2 L^{-1} , respectively, comply with this resolution, however, the resolution emphasises that the established limit can only be exceeded in the case of effluent from a treatment system with a minimum BOD removal efficiency of 60%, or by means of a self-depuration study of the water body which proves that the targets of the framework of the receiving body have been met, in the case of the system in question. According to Mendonça et al. (2005) in their studies, the Roza Elze WWTP scored an average efficiency of 79%, thus meeting the average results obtained of 137.7 mg o2 L^{-1} in the experiment.

According to Duarte (2006), hydrogen ion concentration is an important parameter for analysing water quality, whether it is surface water or wastewater. The pH values that promote the existence of most biological organisms are between 6 and 9 (CONAMA, 2005). In view of this, wastewater with a high concentration of the H ion$^+$, i.e. low pH, is difficult to treat biologically. It should be emphasised that the importance of pH is not only in the biological and chemical reactions that take place in sewage treatment, but when the water is used for irrigation, the fact that the pH is too acidic or too basic can lead to serious nutrition and toxicity problems for the crop, as well as the appearance of incrustations and even corrosion in the irrigation systems (DUARTE, 2006).

Duarte (2006) apud Ayres & Westcot (1991) points out that in relation to the effect of pH on water intended for irrigation, values between 6.5 and 8.4 are recommended. The concentration of H^+ and OH^- in irrigation water can influence the availability and absorption of nutrients by plants, the structure and properties of the soil and the irrigation system. Therefore, according to the data presented in Tables 4.2 and 4.3, all types of water are within the range considered ideal by the authors cited and do not represent negative effects in terms of irrigation practice.

According to Feigin et al. (1991), wastewater is an important source of phosphorus and, when used in irrigation, restores sources of this nutrient in the soil. However, excess amounts of this nutrient can lead to an induced deficiency of copper, zinc and iron, making it necessary to apply these micronutrients to the soil or plant leaves. However, it is important to note that a considerable source of phosphate fertiliser is of interest for agricultural reuse.

According to CONAMA Resolution No.° 357/2005, the maximum total phosphorus content for lentic environments should be no more than 0.020 mg L^{-1} , so the first sample of treated water was not within the parameter, but the second sample was acceptable. Kouraa et al. (2002), when reusing wastewater treated by a series of anaerobic, aerated and facultative ponds for irrigating potatoes and lettuce, quantified an average value of total phosphorus corresponding to 2.77 mg L^{-1} and, at the end of the cycle, found that there was no significant difference in the physical-chemical parameters of the soil after irrigation. The average of the total phosphorus results in this study was 2.54 mg L^{-1} , lower than the average obtained in the study mentioned above, as well as the results of all the samples with wastewater, whose highest value was AR 03 at 2.65 mg L^{-1} .

According to Duarte (2006), the main sources of nitrogen in receiving bodies and surface waters come from the discharge of domestic and industrial waste, human excreta and synthetic fertilisers. Nitrogen is an important nutrient for plant growth and, when present in water intended for irrigation, has the effect of nitrogen used as a fertiliser. However, high amounts of the element can lead to disorganised growth and delayed fruit ripening, resulting in low quality harvests.

According to the results shown in Table 4.3 and 4.4, for the first sample with treated water with a pH of less than 7.5, the maximum permissible value of total ammoniacal nitrogen for class 1 is 3.7 mg L^{-1} and therefore meets the established parameters with a value of 0.3942 mg L^{-1} , in the second sample the pH is between 7.5 and 8.0 and therefore the parameter is 2.0 mg L^{-1} and the result was not detected. Duarte (2006) points out in his studies that irrigation with this type of water showed an average ammoniacal nitrogen value of 25.41 mg L^{-1} for growing chilli peppers, a fact that contributed to the increase in crop production. However, the author emphasises that the nitrate content in the soil must be taken into account, as it poses an imminent risk of pollution to groundwater. Therefore, in the case of wastewater, an average of 22.20 mg L^{-1} was obtained in this study, a value that does not exceed those mentioned above, as well as the results obtained in all the samples with wastewater, the highest value of which was AR 06 at 24.44 mg L^{-1} .

4.4. ANALYSIS OF THE MICROBIOLOGICAL QUALITY OF SUNFLOWER DRY MATTER

In the reuse of wastewater for irrigation, the contaminants of importance to public health are biological. Determining the amount of pathogenic organisms present in wastewater is of paramount importance due to the high risk that its use can pose to public health. Consideration of risk management implies that guidelines are not produced with the aim of being applied directly and absolutely in all countries; in fact, guidelines aim to establish a certain level of public health in relation to pre-established risks, thus providing a common reference for establishing national or regional standards.

The microbiological results obtained in this study were compared with established federal legal standards. It should be pointed out that there is no legislation on animal feed in the form of silage, so this study used the legislation on human food (Agência Nacional de Vigilância Sanitária - ANVISA), which in a way makes the analysis more rigorous. It can be seen that all the treatments used were within the ANVISA standards, so treatment five (T5), composed entirely of treated wastewater, should be recommended.

The mention of a count < 3.0 for the conventional methodology indicates that none of the tubes inoculated were positive, and can be seen as an absence of faecal coliforms per g or ml of product. Resolution RDC nº 12 of 02/01/2001 (ANVISA, 2001) establishes a maximum faecal coliform count (thermotolerant coliforms or E. coli) for flours, pasta and similar, in the sub-item of starch-based products, semi-manufactured flours (industrialised processes) stable at room temperature of 50 $MPN.g^{-1}$. As explained above and shown in Table 4.4, the results found were < 3.0 $NMP.g^{-1}$, so it can be seen that they meet the parameters of the legislation. In relation to the Salmonella sp. analysis, the recommended microbiological standards are the absence of Salmonella sp. in 25 grams, values which are duly met, as also shown in Table 4.4, due to the fact that, for the preparation of the silage, the material was placed in an oven at 65° C and these bacteria denature to death at this temperature.

Table 4.4 - Microbiological results of the ground sunflower aerial part

Treatment	Thermotolerant coliforms (MPN g $)^{-1}$	E. *coli* (CFU g $)^{-1}$	Moulds and yeasts (CFU g $)^{-1}$	*Salmonella sp.* (in 25 g)	Proportions
T1R1	<3,0	<3,0	9	Absence	
T1R2	<3,0	<3,0	9	Absence	100% water
T1R3	<3,0	<3,0	<10	Absence	DESO
T1R4	<3,0	<3,0	<10	Absence	
T2R1	<3,0	<3,0	<10	Absence	
T2R2	<3,0	<3,0	<10	Absence	100% water
T2R3	<3,0	<3,0	9	Absence	waste
T2R4	<3,0	<3,0	<10	Absence	
T3R1	<3,0	<3,0	<10	Absence	
T3R2	<3,0	<3,0	<10	Absence	50% water
T3R3	<3,0	<3,0	<10	Absence	DESO/
T3R4	<3,0	<3,0	<10	Absence	waste water
T4R1	<3,0	<3,0	9	Absence	25% water
T4R2	<3,0	<3,0	<10	Absence	DESO
T4R3	<3,0	<3,0	<10	Absence	75% water
T4R4	<3,0	<3,0	<10	Absence	waste
T5R1	<3,0	<3,0	<100	Absence	25% water
T5R2	<3,0	<3,0	<10	Absence	waste
T5R3	<3,0	<3,0	<10	Absence	75% water
T5R4	<3,0	<3,0	<10	Absence	DESO

Source: ITPS (2012)

The results obtained in the present study corroborate the observations made by Al-Nakshabandi et al. (1997) and Emongor (2004), when these authors found the absence of faecal coliforms, Salmonella sp., Shigella sp. and E. coli in all the analysed samples of aubergine and tomatoes irrigated with treated wastewater. Koraa et al. (2002) also found that crops irrigated with treated wastewater and drinking water had faecal coliform concentrations of less than 3.0 and helminth eggs of less than 1 egg.L^{-1} .

With regard to moulds and yeasts, the microbiological standards recommended for specific population groups were also used, with values of 50 CFU.g^{-1} , which, when compared with Table 4.5, shows that the values obtained were less than 10, with the exception of one repetition in the treatment of 25% wastewater and 75% DESO water, which showed a result of less than 100, not implying that the established standards were not met.

Bastos & Mara (1993) point out that in addition to contamination via irrigation, contamination of crops through contact with the soil occurs because bacteria tend to survive longer in the soil than on crops, since the soil has conditions such as organic matter, humidity, temperature and exposure to the sun that are more favourable for the survival and growth of these organisms. In addition to contact with the soil through harvesting or splashes during irrigation, irrigation management and transport can also lead to contamination of irrigated products.

According to Feachen et al. (1983), viruses and bacteria cannot penetrate plant tissue unless it is damaged. However, some types of pathogenic microorganisms can be present on the surface of crops when they are irrigated or fertilised with organic products. The authors also emphasise that the survival time of pathogenic microorganisms on crops is much shorter than in other types of environments, such as viruses, bacteria and helminths, which survive for a little more than 15 to 60 days, depending on the environment.

Mendonça et al. (2005) point out in their studies that the sewage treatment plant in question has a high detention time for the system, around 141 days, resulting in the absence of parasites in the treated wastewater, showing the high removal capacity of the system studied, due to the fact that these parasites are deposited at the bottom of the stabilisation pond system throughout the process.

In view of the above, treated wastewater can be used to irrigate sunflower crops. It is also important to emphasise that efficient treatment of the effluent to be used, appropriate choice and management of the irrigation system, restrictions on the type of crop to be irrigated and care in harvesting, transport and handling should always be promoted. Attention should be paid to the fact that the accumulated knowledge on the agricultural utilisation of effluents from WWTPs in Brazil is still in its infancy, which makes research and action towards controlled reuse essential, including regulation, because failure to adopt these criteria could lead to the indiscriminate use of treated wastewater for irrigating various crops, thus becoming a major vector for the spread of environmental pollution and water-borne diseases, as pointed out in studies carried out in Mexico and Pakistan by Alvarez (1997) and Van der Hoek et al. (2002), respectively.

It should be noted that the use of sanitary sewage in agriculture depends on joint action by the federal, state and municipal governments, in terms of appropriate planning for land use and occupation, the implementation of infrastructure for collecting and treating the sewage generated and the development of programmes to encourage the use of treated sewage for irrigation. A careful reuse policy transforms the polluting and aggressive problem of sewage into an economic resource, and with proper care and overcoming cultural resistance, it will present itself as a solution that is safe for health, economically viable and environmentally sustainable.

Microbiological quality standards for treated wastewater intended for irrigation will only be definitively established after numerous demonstrations of their sufficiency as a health protection measure. This will be validated through tests under different conditions, such as climate, irrigated crops, irrigation methods and effluent quality. Conclusive evidence of disease transmission (real health risks) can only be obtained through complex epidemiological studies and, as such, the assessment of potential risks is a valuable tool. It is important to emphasise the need for studies aimed at analysing the microbiological quality of the soil used for cultivation, in order to ensure the soil-water-plant system.

This study recommends planting the crop in the field in order to assess its development, given that its development in pots may have been compromised due to the confinement, as well as assessing the contamination rates obtained in the soil, and that state and federal governments begin management processes to establish political, legal and institutional bases for reuse, both for the use of effluents and for national or state water resources plans (PNRH and PERH).

CHAPTER 5

5. CONCLUSIONS

According to the results of this research, it was possible to conclude that:

1. The climate of the region studied suited the requirements of the sunflower crop in terms of solar radiation, relative humidity, maximum, minimum and average temperatures, all of which are found in the semi-arid region of the state, so in climatic terms it is possible to obtain good results throughout the region.
2. The dry matter from the aerial part of ground sunflower under the conditions studied could be used for animal feed, since the results were within acceptable health standards even for human food, since there is no legislation on animal feed, and the influence of irrigation with wastewater has been shown to be positive in various studies on the subject, including increasing the productivity of the crops studied.
3. As there was no substantial relative difference between the different types of treatment used in the research, treatment with 100% of the treated wastewater could be used, optimising the destination of this material.

REFERENCES

NATIONAL WATER AGENCY - ANA. Water is life and life is not wasted" campaign. Available at <http://www.ana.gov.br/SalaImprensa/aguavida/aguavida.asp>. Accessed on: 22 August 2012.

ALLEN, R. G.; PEREIRA, L. S.; RAES, D.; SMITH. M. *Crop evapotranspiration - Guidelines for computing crop water requirements.* In: FAO Irrigation and Drainage Paper 56. Rome: FAO, 1998.

AL-NAKSHABANDI, G. A.; SAQQAR, M. M.; SHATANAWI. M. R.; FAYYAD, M.; AL-HORANI, H. *Some enviromental problems associated with the use treated wastewater for irrigation in Jordan.* Agricultural Water Management, Amsterdam, v. 34, 0.81-94, 1997.

ALVAREZ, H. R. *El Valle de Mezquital*, México: Estudio *de* caso VII. Available at: http://www.cepis.ops-oms.org/eswww/proyecto/repidisc/publica/repindex/repi066/vallemez.html Accessed on 09 January 2013.

AMBROSANO, E. J.; TANAKA, R. T.; MASCARENHAS, H. A. A.; RAIJ, B. van;

QUAGGIO, J. A. & CANTARELLA, H. Legumes and oilseeds. In: RAIJ, B. van; CANTARELLA, H.; QUAGGIO, J. A. & FURLANI, A. M. C., eds.
Fertilisation and liming recommendations for the state of São Paulo. Campinas, Agronomic Institute of Campinas, 1996. p. 187-203 (Technical Bulletin, 100).

BARNI, N. A. Sunflower growth, development and yield models as a function of solar radiation, temperature and water availability. 1994. 249 f. Thesis (Doctorate in Agronomy/Phytotechny-Agrometeorology)- Faculty of Agronomy, UFRGS, Porto Alegre, 1994.

BARTONE, C. R.; ARLOSOROFF, S. *Reuse of pond effluent in developing countries.* Watersci. Technol.19(12): 1987. p. 289-297.

BASTOS, R. K. X.; BEVILACQUA, P. D.; KELLER, R. Pathogenic organisms and effects on human health. In: Disinfection of sanitary effluents. PROSAB 3. Rio de Janeiro: ABES. p. 27-88, 2003.

BASTOS, R. K. X.; MARA, D. D. Evaluation of criteria and standards for the microbiological quality of sanitary sewage with a view to its use in agriculture. In: CONGRESSO BRASILEIRO DE ENGENHARIA SANITÁRIA E AMBIENTAL, 17, 1993, Natal, RN. Proceedings... Natal: UFRN, 1993.

BERGAMASCHI, H. Development of water deficit in crops. In: BERGAMASCHI, H. (Coord.) Agrometeorologia aplicada à irrigação. Porto Alegre: Ed. Universidade/UFRGS, 1992. P. 25-32.

BERNARDO, S.; SOARES, A. A.; MANTOVANI, E. C. Irrigation Manual. 9ª ed. Viçosa: UFV, University Press, 2009.

BLUM, J. R. C. Water Quality Criteria and Standards. In: MANCUSO, P. C. S; SANTOS, H. F. (Ed). Water reuse. Barueri, SP: Malone, 2003. p. 125 - 126.

BRAZIL. Law nº 9.433, of 8 January 1997. Institutes the National Water Resources Policy, Creates the National Water Resources Management System, and Makes Other Provisions. Federal Official Gazette, 09/01/1997.
Available at: http://www.planalto.gov.br/ccivil 03/leis/L9433.htm. Accessed on: 16 June 2012.

BRAZIL. Ministry of Health (MS) National Health Agency (ANVISA). Collegiate Board Resolution (RDC) no. 12, of 2nd January 2001.
Technical Regulation on Microbiological Standards for Food. Official Gazette of the Federative Republic of Brazil Brasília, DF, 10 Feb. 2001. Section 1.
Available at: www.anvisa. gov.br/legis/resol/12 01 rdc.htm. Accessed on: 03 October 2012.

BRAZIL. Bill Nº 5296 of 23 May 2005. Establishes guidelines for public basic sanitation services and the National Basic Sanitation Policy - PNS. National Congress, Brasília, 23/05/2005. Available at:
http://www.camara.gov.br/proposicoesWeb/fichadetramitacao?idProposicao=286716. Accessed on: 16 June 2012.

. Resolution N°54 of 28 November 2005. Establishes modalities, guidelines and general criteria for the practice of non-potable direct reuse and makes other provisions. Diário Oficial da União, Brasília, 09/03/2006. Available at: http://www.cnrhsrh.gov.br/. Accessed on: 16 June 2012.

BROWN, L. R. Eco-Economics: building an economy for the Earth .
Salvador: UMA, 2003.

CÂMARA, J. B. D.; SANTOS, T. C. C. (Orgs). GEO Brasil 2002: perspectives on the environment in Brazil. Brasília: IBAMA, 2002.

CAMARGO, A. L. B. Sustainable Development: dimensions and challenges.
Papirus. Campinas, SP: 2003.

CARDOSO, M. R. A. Environmental Epidemiology. In: PHILLIPI JR (ed.).
Sanitation, health and the environment: foundations for sustainable development. Barueri: Malone, 2005. p. 87-116.

CASTRO, C.; FARIAS, J. R. B. Sunflower ecophysiology. In: LEITE, R. M. V. B. C.; BRIGHENTI, A. M.; CASTRO, C. (Ed.). Sunflower in Brazil. Londrina: Embrapa Soja, 2005 p. 163-218.

CONAMA - NATIONAL ENVIRONMENTAL COUNCIL. RESOLUTION N° 357 of 29 August 2006. Defines criteria and procedures for the agricultural use of sewage sludge generated in sewage treatment plants and its derived products, and makes other provisions. Ministry of Urban Development and the Environment, 2006. Available at: www.mma.gov.br/port/conama/legiabre. cfm?codlegi=506. Accessed on: 03 December 2012.

CONAMA - NATIONAL ENVIRONMENTAL COUNCIL. RESOLUTION N° 430 of 13 May 2011. Provides for the conditions and standards for effluent discharge, complements and amends Resolution n° 357, of 17 March 2005, of the National Environment Council-CONAMA. Ministry of Urban Development and the Environment, 2006. Available at: www.mma.gov.br/port/conama/legiabre.cfm?codlegi=646. Accessed on: 14 January 2013.

CONNOR, D. J. & Hall, A. J. (1997). Sunflower physiology. In: A.A. Schneiter. Sunflower technology and production (pp. 113-183). Agron. Monogr. 35. ASA, CSSA and SSSA, Madison, WI.

CONWAY, G. Food production in the 21st century: biotechnology and the environment. Trad. Celso Mauro Paciornik. - São Paulo: Estação Liberdade, 2003.

CROOK, J. Water reclamation and reuse criteria. In: ASANO, T. Wastewater reclamation and reuse. Lancaster: Technomic Publishing, 1998. p. 627-704.

DALTRO FILHO, Current Situation of Stabilisation Ponds in the State of Sergipe - Brazil, 1997.

DI BERNARDO, L.; DI BERNARDO, A.; CENTURIONE FILHO, P. L. Treatability tests for water and waste generated in water treatment plants. São Carlos: Editora Rima, 2002.

NATIONAL DEPARTMENT OF WATER AND ELECTRICITY - GENERAL COORDINATION OF WATER RESOURCES. Map of water availability in Brazil. Brasilia, 1992.

DOORENBOS, J.; PRUITT, W. O. Water Requirements of Crops. 5 ed., Rome: FAO, 1977. 204 p. (FAO Studies, Irrigation and Drainage, 24).

DOWNES, F. P.; ITO, K. (Eds.). Compendium of methods for the microbiological examination of foods. Washington: Apha, 2001.

DUARTE, A. Reuse of treated wastewater in the irrigation of pepper crops. 2006. 188 f. Thesis (Doctorate in Agronomy). University of São Paulo, Piracicaba, 2006.

EIGER, S. Self-depuration of watercourses. In: MANCUSO, P. C. S.; SANTOS H. F. (Eds). Water reuse. Barueri, SP: Malone, 2003. p 233 - 259.

EMBRAPA. Sunflower cultivation. In: CASTRO, César de. Technical Circular° 13, 1997.

EMONGOR, V. E.; RAMOLEMANA, G. M. Treated sewage effluent (water) potential to be used for horticultural production in Botswana. Physics and chemistry of the earth, Oxford, v.29, p.1101-1108, 2004.

ENNES, M. A. Identity, nature and sustainability. In: SANTOS, Antônio Carlos dos Santos. Philosophy & Nature. São Cristóvão, SE: EdUFS, 2008.

EVANGELISTA, A. R.; LIMA, J. A. Utilisation of sunflower silage in animal feed. Anais do Simpósio Sobre Produção e Utilização de Forragens Conservadas / Editors Clóves Cabreira Jobim, Ulysses Cecato, Júlio César Damasceno e Geraldo Tadeu dos Santos. - Maringá : UEM/CCA/DZO, p. 177-217, 2001.

FACCIOLI, G. G. et al. Basic irrigation management in vegetable production. Easy Technology Collection No. 18. 1ª edition. LK. Brasília: 2006.

FAO, UN, Annual Report, 2007.

FEACHEM, R. G.; BRADLEY, D. J.; GARELICK, H.; MARA, D. D. Sanitation and disease: health aspects of excreta and wastewater management. Chichester: John Wiley, 1983a.

FEIGIN, A.; RAVINA, I; SHALHEVET, J. Irrigation with treated sewage effluent: management for environmental protection. Berlin: Springer-Verlag, 1991.

FERREIRA, A. M. D. R. P. Microclimate and sunflower development in semi-arid Mediterranean conditions. 2007. 212 f. Thesis (Doctorate in Agronomic Engineering) - Instituto Superior de Agronomia. Lisbon: 2007.

FOOD AND AGRICULTURE ORGANISATION. Rome. Wastewater treatment and use in agriculture. FAO Study: Water and Drainage No. 47, 1992.

FRANCO, B. D. G. M.; LANDGRAF, M. Microbiologia de alimentos. São Paulo: Atheneu, 1996.

GALBIATTI, J. A.; CAVALCANTE, Í. H. L.; RIBEIRO, A. G.; BECKMANN-Cavalcante, M. Z. Fertilisation and irrigation water quality on lettuce growth and development. Scientia Agraria, v.8, n.2, p.185-192, 2007.

HELDWEIN, A. B.; MALDANER, I. C.; RADONS, S. Z.; LOOSE, L. H.; LUCAS, D. D. P.; HINNAH, F. D. Estimation of radiation balance in sunflower as a function of global solar radiation. Brazilian Journal of Agricultural and Environmental Engineering, 2012.

HESPANHOL, I. Potential for water reuse in Brazil: agriculture, industry, municipalities and aquifer recharge. In: MANCUSO, P. C. S. & SANTOS, H. F. (editors). Water reuse. Barueri-SP: Manole, p.37-95, 2003.

HESPANHOL, I. Public health and agricultural reuse of sewage and biosolids. In: Mancuso, P. C. S.; Santos, H. F. dos (ed.). Water reuse. Barueri: Manole, 2003. p.97-124.

HESPANHOL, I. A new paradigm for water resources management. Estudos Avançados, v.22, n.63, p.131-158, 2008.

HESPANHOL, I.; PROST, A. Who guidelines and national standards for reuse and water quality. Water Research, 28 (1). London, 1994.

HIRATA, R. Water Resources. In: TEIXEIRA, W. et al. Deciphering the Earth. São Paulo: Oficina de Textos, 2000.

ISO 6887-1. *Microbiology of food and animal feeding stuffs - Preparation of test samples, initial suspension and decimal dilutions for microbiological examination p Part 1: General rules for the preparation of the initial suspension and decimal dilutions*, 1st ed. The International Organization for Standardization, 1999.

JAY, J. M. *Modern food microbiology*. 5 ed. Gaithersburg. Aspen Publishers, 1998.

KORAA, A.; FETHI, F.; LAHLOU, A.; OUAZZANI N. *Reuse of urban wastewater by combined stabilisation pond system in Benslimane (Marocco).*
Urban Water, Amsterdam, v.4, p.373-378, 2002.

LEEUWEN, J. Van. *Reclaimed water - an untapped resourse. Desalination,* n. 106. 1995.

LEITE, R. M. V. B. C.; BRIGHENTI, A. M.; CASTRO, C. de. (Ed.). Sunflower in Brazil. Londrina: Embrapa Soja, 2005.

LUCAS FILHO, M.; ANDRADE NETO, C. O.; MELO, H. N. S.; PEREIRA, M. G. Evolution of the treated sewage disposal process through sub-surface runoff in soil prepared with vegetation cover. In: Post-treatment of effluents from anaerobic reactors. Collection of technical papers - Volume 2.
PROSAB/FINEP Project - Rio de Janeiro, 2001.

MANCUSO, P. C. S.; SANTOS, H. F. dos S. (eds). Water Reuse. Barueri: University of São Paulo, School of Public Health, Environmental Health Information Centre, 2003.

MANTOVANI, E. C; BERNARDO, S.; PALARETTI, L. F. P. Irrigation - principles and methods. Viçosa: UFV, University Press, 2009.

MARQUES, M. O. et al. Use of treated sewage in irrigation: agronomic and environmental aspects. In: BASTOS, R. K. X. (coord.). Use of treated sewage in fertigation, hydroponics and fish farming. Rio de Janeiro: ABES, RIMA, 2003. p. 61-116

MARTINEZ, H. E. P.; FILHO, J. B. da S. Introduction to hydroponic plant cultivation. Viçosa: UFV, 1997.

MASTNY, L.; CINCOTTA, R. P. Analysing links between population and security. In: LOPES, C (Presentation); MULLETT, H.; MULLETT C. (Trad.). State of the World, 2005: State of Consumption and Sustainable Consumption. Wordlwatch Institute. Salvador: UMA, 2005.

MENDONÇA, J.C. et al. Comparison between methods for estimating reference evapotranspiration (Et0) in the northern region of Rio de Janeiro State. Revista Brasileira de Engenharia Agrícola e Ambiental, v.7, n.2, p.275-279, 2003 Campina Grande, PB,

DEAg/UFCG.

MENDONÇA, L. C.; PINTO, A. S.; SAMPAIO, L. F. S.; CARDOSO, L. R. Caracterização e Avaliação da ETE Rosa Elze para Reuso do Efluente, Revista Brasileira de Engenharia Agrícola e Ambiental, v. 9 supl, p.143-145, 2005.

METCALF; EDDY. Wastewater Engineering: treatment, disposal and reuse. New York: McGraw-Hill, 2003.

MONTEITH, J. L. The heat balance of soil beneath crops. In: CLIMATOLOGY and Microclimatology. Paris: UNESCO, 1958.

MORTON, D. R. Aerobic plate count. In: DOWNES, F.P.; ITO, K. Compendium of methods for the microbiological examination of foods. 4.ed. Washington: American Public Health Association, 2001. Chap.37, p.63-67.

MORUZZI, R. B. Water reuse in the context of water resources management: impact, technologies and challenges. Rio Claro, SP, 2008.

MOTA, S.; BEZERRA, F. C.; TOMÉ, L. M. Evaluation of the performance of crops irrigated with treated sewage. Brazilian Association of Sanitary and Environmental Engineering-ABES. Fortaleza-CE, 2000.

UN - Available at: http://www.onu.org.br/a-onu-em-acao/a-onu-e-o-meio-environment/print/. Accessed on: 10 June 2012.

UN - United Nations Organisation. Water and sanitation facts. United Nations Department of Public Information, June 2012. Available at: http://www.onu.org.br/rio20/agua.pdf. Accessed on: 16 June 2012.

PAGANINI, W. S. Disposal of sewage on the ground (surface runoff). São Paulo: Fundo Editorial da AESABESP, 1997.

PAGANINI, W. S. Water reuse in agriculture. In:. MANCUSO P. C. S., SANTOS H. F. (Eds). Water reuse. Baureri, SP: Manole, 2003.

PELCZAR, JR, M. J.; CHAN, E. C. S.; KRIEG, N. R. Microbiologia: conceitos e aplicações. 2ed. São Paulo: McGraw-Hill, 1997. V.2. ch. 30, p. 372-397: Food Microbiology.

PESCOD, M. B. Wastewater treatment and use in agriculture. Rome, 1992. 125p. (FAO irrigation and drainage paper, 47).

PESCOD, M. B.; ALKA, U. Guidelines for wasterwater reuse in agricultural. In: Regional seminar on the treatment and use of sewage, effluent for irrigation, Food And Agriculture. Rome: Organisation of United Nations. Nicosia, 1985.

UNITED NATIONS ENVIRONMENT PROGRAMME. World Environment Outlook 2002: Geo 3, past present and future.
OSTORINO, R. (coord.); SHELLARD, S.; CORREA, N. B (Trads). Brasília: IBAMA/UMA: 2004.

PRUITT, W. O.; von OETTINGEN, S.; MORGAN, D. L. Central California evapotranspiration frequencies. Journal of the Irrigation and Drainage Division, New York, v. 98, n. IR- 2, 1972.

RAWSON, H. M.; CONSTABLE, G. A.; HOWE, G. N. Carbon production of sunflower cultivars in field and controlled environments. II. Leaf growth. Australian Journal of Plant Physiology, Melbourne, v.7, p. 575-586, 1980.

RIJSBERMAN, F. R. Water scarcity: Fact or fiction? Agricultural Water Management, v. 80, p.5-22, 2006.

RODRIGUES, R. S. As dimensões legais e institucionais do reuso de água no Brasil: proposta de regulamentação do reuso no Brasil. São Paulo: Polytechnic School of the University of São Paulo, 2005.

SACHS, I. Development: inclusive, sustainable and sustained. Garamond. Rio de Janeiro: 2008.

SACHS, I. Ecodevelopment: growth without destruction. São Paulo: Vértice, 1986.

SEMENTES CONTIBRASIL. Sunflower: producer's manual. Campinas-SP, 1981.

SHUVAL, H. I., ADIN, A., FATTAL, B., RAWITZ, E. and YEKUTIEL, P. Wastewater Irrigation in Developing Countries: Health Effects and Technological Solutions. World Bank Technical Paper Number 51. The World Bank, Washington, DC. 1986

SIEBE, C. Heavy metal availability to plants in soil irrigated with wastewater from México city. Wat. Sci.Tech. Elsevier, vol.32, n. 12, p. 29, 1996.

SILVA, N.; JUNQUEIRA, V. C. A.; SILVEIRA, N. F. A.; TANIWAKI, M. H.; SANTOS, R. F. S.; GOMES, R. A. R. Manual of Methods for Microbiological Analysis of Food and Water. 4ª edition. São Paulo. Varela Publishing House, 2010.

SIQUEIRA, R. S. Manual of food microbiology. Brasília: EMBRAPA, SPI; Rio de Janeiro EMBRAPA, CTAA, 1995.

SOUZA, A.; OLIVEIRA. M. F.; CASTIGLIONI. V. B. R. Boron in sunflower cultivation. Londrina, PR. Article 2004.

SPERLING, M. V. Introduction to water quality and sewage treatment. - 3.ed.- Belo Horizonte: Department of Sanitary and Environmental Engineering; Federal University of Minas Gerais, 2005.

SPERLING, M. V. Princípios do tratamento biológico de águas residuárias: Lagoas de Estabilização. 2. ed. Belo Horizonte: Ed. UFMG, 2002.

STRAUSS, M.; HEINSS, U.; MONTANGERO, A. On-site sanitation: when the pits are *full pplnnnigf for resource protection in faecal sludge management. In: Proceedings, International Conference on Resolving Conflicts between drinking WatorSemand acdPreseures from Society's (CasOes* (Chorus et al., editors). Bad Elster, Germany, 24-28 November, 2000.

SWANSON, H. M. J.; BUSTA, F. F.; PETERSON, E. H.; JOHNSON, M. G. Colony count methods. In: VANDERZANT, V.; SPLITTSTOESSER, D. F. (Ed.) *Compendium of methods for the microbiological examination offoods*. 3 ed. Washington: American Public Health Association, 1992. Chap. 4, p. 75-95.

U.S. GEOLOGICAL SURVEY-USGS. *The water cycle*. Available at: <http://ga.water.usgs.gov/edu/watercycleportuguese.html>. Accessed on: 10 June 2012.

UNGARO, M. R. G.; CASTRO, C.; FARIAS, J.R.B.; BARNI, N. A.; RAMOS, N.P.; SENTELHAS, P. C. Org. MONTEIRO. J. E. B. A. Agrometeorologia dos cultivos: o fator meteorológico na produção agrícola. Brasília: INMET, 2009.

Van der HOEK, W.; HASSAM, M.U.; ENSINK, J.H.J.; FEENSTRA, D.; RASCHID-SALLY, L.; MUNIR, S.; ASLAM, R.; ALI, N.; HUSSAIN, R.; MATSUNO, Y. *Urban wasterwater: a valuable resource for agriculture: a case study from Haroonabad, Pakistan.* Available at:
http://www.iwmi.cgiar.org/health/wastew/index.htm. Accessed on 09 January 2013.

VAZQUEZ-MONTIEL, O.; HORAN, N. J.; MARA, D. D. *Management of domestic wastewater for reuse in irrigation.* Wat. Sci. Tech. Great Britain: Elsevier Science, v. 33, n.10-11, p. 355-362, 1996.

VEIGA, J. E. da. Rethinking development. In:. Environment & development. São Paulo: Senac, 2006. p. 119-180.

WANDERLEY, T. F. Evaluation of the Effects of Sewage Water Reuse on the Productivity and Microbiological Quality of Sweet Potato Cultivars for Biomass Production. 2005. 130 f. Dissertation (Master's Degree in Environmental Sciences) - Federal University of Tocantins. Tocantins: 2005.

WHO, Health guidelines for the use of wastewater in agriculture and aquaculture. Technical Report Series, No. 776, 1989.

WHO, Reuse of effluents: methods of wastewater treatment and health safeguards. Report of a WHO Meeting of Experts. Geneva, World Health Organisation (Technical Report Series Nº . 517), 1973.

WHO, Wastewater Use in Agriculture, in: Guidelines for the Safe Use of Wastewater, Excreta and Greywater, vol.2, World Health Organisation, Geneva, Switzerland, 2006.

WOLF, A. T., KRAMER, A.; CARIUS, A.; DABELKO, G. D. Managing Water Disputes and Co-operation. In: MULLETT,H.; MULLETT, C. (transl.) State of the World, 2005: State of Consumption and Sustainable Consumption. Wordlwatch Institute. Salvador: UMA, 2005.

ANNEXES:

INSTITUTO TECNOLÓGICO E DE PESQUISAS DO ESTADO DE SERGIPE

Rua Campo do Brito, Nº371, Treze de Julho, CEP 49.020-380
Aracaju - SE - Brasil

Fone (79) 3179-8081/8087 Fax (79) 3179-8087/8090
CNPJ 07.258.529/0001-59

Relatório de Ensaios ITPS Nº 3483/12

Revisão 03

Cliente	LUCIANA COELHO MENDONÇA	Telefone	
Endereço	RUA FRANCISCO RABELO LEITE NETO, 990 CASA 8	Contato(s)	8817-9580
e-mail	lumendon@uol.com.br	Fax	
Amostra(s)	Rações	Recepção	03/10/12

Amostra	1			Código 3483/12-01	Coleta em
Ensaio	**Resultado**	**Unidade**	**LQ**	**Método**	**Data do Ensaio**
Coliformes a 45°C	<3,0	NMP/g	–	SMEWW9221B	03/10/12
Escherichia Coli	<3,0	UFC/g	–	Cont. em Placas	03/10/12
Bolores e Leveduras em 25g	<1,0 X 10^2	UFC/g	–	Cont. em Placas	03/10/12
Salmonelas	Ausência	em 25g	–	AOACC 967.26	03/10/12

Amostra	2			Código 3483/12-02	Coleta em
Ensaio	**Resultado**	**Unidade**	**LQ**	**Método**	**Data do Ensaio**
Coliformes a 45°C	<3,0	NMP/g	–	SMEWW9221B	03/10/12
Escherichia Coli	<3,0	UFC/g	–	Cont. em Placas	03/10/12
Bolores e Leveduras em 25g	<1,0 X 10	UFC/g	–	Cont. em Placas	03/10/12
Salmonelas	Ausência	em 25g	–	AOACC 967.26	03/10/12

Amostra	3			Código 3483/12-03	Coleta em
Ensaio	**Resultado**	**Unidade**	**LQ**	**Método**	**Data do Ensaio**
Coliformes a 45°C	<3,0	NMP/g	–	SMEWW9221B	03/10/12
Escherichia Coli	<3,0	UFC/g	–	Cont. em Placas	03/10/12
Bolores e Leveduras em 25g	<1,0 X 10	UFC/g	–	Cont. em Placas	03/10/12
Salmonelas	Ausência	em 25g	–	AOACC 967.26	03/10/12

Amostra	4			Código 3483/12-04	Coleta em
Ensaio	**Resultado**	**Unidade**	**LQ**	**Método**	**Data do Ensaio**
Coliformes a 45°C	<3,0	NMP/g	–	SMEWW9221B	03/10/12
Escherichia Coli	<3,0	UFC/g	–	Cont. em Placas	03/10/12
Bolores e Leveduras em 25g	9	UFC/g	–	Cont. em Placas	03/10/12
Salmonelas	Ausência	em 25g	–	AOACC 967.26	03/10/12

Amostra	5			Código 3483/12-05	Coleta em
Ensaio	**Resultado**	**Unidade**	**LQ**	**Método**	**Data do Ensaio**
Coliformes a 45°C	<3,0	NMP/g	–	SMEWW9221B	03/10/12
Escherichia Coli	<3,0	UFC/g	–	Cont. em Placas	03/10/12
Bolores e Leveduras em 25g	<1,0 X 10	UFC/g	–	Cont. em Placas	03/10/12
Salmonelas	Ausência	em 25g	–	AOACC 967.26	03/10/12

Amostra	6			Código 3483/12-06	Coleta em
Ensaio	**Resultado**	**Unidade**	**LQ**	**Método**	**Data do Ensaio**
Coliformes a 45°C	<3,0	NMP/g	–	SMEWW9221B	03/10/12
Escherichia Coli	<3,0	UFC/g	–	Cont. em Placas	03/10/12
Bolores e Leveduras em 25g	<1,0 X 10	UFC/g	–	Cont. em Placas	03/10/12
Salmonelas	Ausência	em 25g	–	AOACC 967.26	03/10/12

Amostra	7			Código 3483/12-07	Coleta em
Ensaio	**Resultado**	**Unidade**	**LQ**	**Método**	**Data do Ensaio**
Coliformes a 45°C	<3,0	NMP/g	–	SMEWW9221B	03/10/12
Escherichia Coli	<3,0	UFC/g	–	Cont. em Placas	03/10/12
Bolores e Leveduras em 25g	<1,0 X 10	UFC/g	–	Cont. em Placas	03/10/12
Salmonelas	Ausência	em 25g	–	AOACC 967.26	03/10/12

A Custodia das amostras é de 15 dias após emissão do relatório de ensaios, exceto para solos que é 90 dias e água que é 2 dias. Não se aplica a amostras perecíveis. Os resultados têm significado restrito e aplicam-se somente às amostras ensaiadas. Este relatório somente poderá ser reproduzido em sua totalidade. O ITPS se isenta de qualquer responsabilidade pela reprodução parcial do mesmo.

RF-LBW-004, Rev. 00 Página: 1/4

INSTITUTO TECNOLÓGICO E DE PESQUISAS DO ESTADO DE SERGIPE

Rua Campo do Brito, Nº371, Treze de Julho, CEP 49.020-380 Aracaju - SE - Brasil
Fone (79) 3179-8081/8087 Fax (79) 3179-8087/8090
CNPJ 07.258.529/0001-59

Relatório de Ensaios ITPS Nº 3483/12

Revisão 03

Cliente	LUCIANA COELHO MENDONÇA	Telefone	
Endereço	RUA FRANCISCO RABELO LEITE NETO, 990 CASA 8	Contato(s)	8817-9580
e-mail	lumendon@uol.com.br	Fax	
Amostra(s)	Rações	Recepção	03/10/12

Amostra	8			Código	3483/12-08	Coleta em	
Ensaio	Resultado	Unidade	LQ	Método		Data do Ensaio	
Coliformes a 45℃	<3,0	NMP/g	–	SMEWW9221B		03/10/12	
Escherichia Coli	<3,0	UFC/g	–	Cont. em Placas		03/10/12	
Bolores e Leveduras em 25g	<1,0 X 10	UFC/g	–	Cont. em Placas		03/10/12	
Salmonelas	Ausência	em 25g	–	AOACC 967.26		03/10/12	

Amostra	9			Código	3483/12-09	Coleta em	
Ensaio	Resultado	Unidade	LQ	Método		Data do Ensaio	
Coliformes a 45℃	<3,0	NMP/g	–	SMEWW9221B		03/10/12	
Escherichia Coli	<3,0	UFC/g	–	Cont. em Placas		03/10/12	
Bolores e Leveduras em 25g	5	UFC/g	–	Cont. em Placas		03/10/12	
Salmonelas	Ausência	em 25g	–	AOACC 967.26		03/10/12	

Amostra	10			Código	3483/12-10	Coleta em	
Ensaio	Resultado	Unidade	LQ	Método		Data do Ensaio	
Coliformes a 45℃	<3,0	NMP/g	–	SMEWW9221B		03/10/12	
Escherichia Coli	<3,0	UFC/g	–	Cont. em Placas		03/10/12	
Bolores e Leveduras em 25g	5	UFC/g	–	Cont. em Placas		03/10/12	
Salmonelas	Ausência	em 25g	–	AOACC 967.26		03/10/12	

Amostra	11			Código	3483/12-11	Coleta em	
Ensaio	Resultado	Unidade	LQ	Método		Data do Ensaio	
Coliformes a 45℃	<3,0	NMP/g	–	SMEWW9221B		03/10/12	
Escherichia Coli	<3,0	UFC/g	–	Cont. em Placas		03/10/12	
Bolores e Leveduras em 25g	<1,0 X 10	UFC/g	–	Cont. em Placas		03/10/12	
Salmonelas	Ausência	em 25g	–	AOACC 967.26		03/10/12	

Amostra	12			Código	3483/12-12	Coleta em	
Ensaio	Resultado	Unidade	LQ	Método		Data do Ensaio	
Coliformes a 45℃	<3,0	NMP/g	–	SMEWW9221B		03/10/12	
Escherichia Coli	<3,0	UFC/g	–	Cont. em Placas		03/10/12	
Bolores e Leveduras em 25g	<1,0 X 10	UFC/g	–	Cont. em Placas		03/10/12	
Salmonelas	Ausência	em 25g	–	AOACC 967.26		03/10/12	

Amostra	13			Código	3483/12-13	Coleta em	
Ensaio	Resultado	Unidade	LQ	Método		Data do Ensaio	
Coliformes a 45℃	<3,0	NMP/g	–	SMEWW9221B		03/10/12	
Escherichia Coli	<3,0	UFC/g	–	Cont. em Placas		03/10/12	
Bolores e Leveduras em 25g	5	UFC/g	–	Cont. em Placas		03/10/12	
Salmonelas	Ausência	em 25g	–	AOACC 967.26		03/10/12	

Amostra	14			Código	3483/12-14	Coleta em	
Ensaio	Resultado	Unidade	LQ	Método		Data do Ensaio	
Coliformes a 45℃	<3,0	NMP/g	–	SMEWW9221B		03/10/12	
Escherichia Coli	<3,0	UFC/g	–	Cont. em Placas		03/10/12	
Bolores e Leveduras em 25g	<1,0 X 10	UFC/g	–	Cont. em Placas		03/10/12	
Salmonelas	Ausência	em 25g	–	AOACC 967.26		03/10/12	

A Custódia das amostras é de 15 dias após emissão do relatório de ensaios, exceto para solos que é 90 dias e água que é 2 dias. Não se aplica a amostras perecíveis. Os resultados têm significado restrito e aplicam-se somente às amostras ensaiadas. Este relatório somente poderá ser reproduzido em sua totalidade. O ITPS se isenta de qualquer responsabilidade pela reprodução parcial do mesmo.

INSTITUTO TECNOLÓGICO E DE PESQUISAS DO ESTADO DE SERGIPE

Rua Campo do Brito, Nº371, Treze de Julho, CEP 49.020-380
Aracaju - SE - Brasil

Fone (79) 3179-8081/8087 Fax (79) 3179-8087/8090
CNPJ 07.258.529/0001-59

Relatório de Ensaios ITPS Nº 3483/12

Revisão 03

Cliente	LUCIANA COELHO MENDONÇA	**Telefone**	
Endereço	RUA FRANCISCO RABELO LEITE NETO, 990 CASA 8	**Contato(s)**	8817-9580
e-mail	lumendon@uol.com.br	**Fax**	
Amostra(s)	Rações	**Recepção**	03/10/12

Amostra	15		Código 3483/12-15	Coleta em	
Ensaio	**Resultado**	**Unidade**	**LQ**	**Método**	**Data do Ensaio**
Coliformes a 45°C	<3,0	NMP/g	–	SMEWW9221B	03/10/12
Escherichia Coli	<3,0	UFC/g	–	Cont. em Placas	03/10/12
Bolores e Leveduras em 25g	<1,0 X 10	UFC/g	–	Cont. em Placas	03/10/12
Salmonellas	Ausência	em 25g	–	AOACC 967.26	03/10/12

Amostra	16		Código 3483/12-16	Coleta em	
Ensaio	**Resultado**	**Unidade**	**LQ**	**Método**	**Data do Ensaio**
Coliformes a 45°C	<3,0	NMP/g	–	SMEWW9221B	03/10/12
Escherichia Coli	<3,0	UFC/g	–	Cont. em Placas	03/10/12
Bolores e Leveduras em 25g	<1,0 X 10	UFC/g	–	Cont. em Placas	03/10/12
Salmonellas	Ausência	em 25g	–	AOACC 967.26	03/10/12

Amostra	17		Código 3483/12-17	Coleta em	
Ensaio	**Resultado**	**Unidade**	**LQ**	**Método**	**Data do Ensaio**
Coliformes a 45°C	<3,0	NMP/g	–	SMEWW9221B	03/10/12
Escherichia Coli	<3,0	UFC/g	–	Cont. em Placas	03/10/12
Bolores e Leveduras em 25g	<1,0 X 10	UFC/g	–	Cont. em Placas	03/10/12
Salmonellas	Ausência	em 25g	–	AOACC 967.26	03/10/12

Amostra	18		Código 3483/12-18	Coleta em	
Ensaio	**Resultado**	**Unidade**	**LQ**	**Método**	**Data do Ensaio**
Coliformes a 45°C	<3,0	NMP/g	–	SMEWW9221B	03/10/12
Escherichia Coli	<3,0	UFC/g	–	Cont. em Placas	03/10/12
Bolores e Leveduras em 25g	<1,0 X 10	UFC/g	–	Cont. em Placas	03/10/12
Salmonellas	Ausência	em 25g	–	AOACC 967.26	03/10/12

Amostra	19		Código 3483/12-19	Coleta em	
Ensaio	**Resultado**	**Unidade**	**LQ**	**Método**	**Data do Ensaio**
Coliformes a 45°C	<3,0	NMP/g	–	SMEWW9221B	03/10/12
Escherichia Coli	<3,0	UFC/g	–	Cont. em Placas	03/10/12
Bolores e Leveduras em 25g	<1,0 X 10	UFC/g	–	Cont. em Placas	03/10/12
Salmonellas	Ausência	em 25g	–	AOACC 967.26	03/10/12

Amostra	20		Código 3483/12-20	Coleta em	
Ensaio	**Resultado**	**Unidade**	**LQ**	**Método**	**Data do Ensaio**
Coliformes a 45°C	<3,0	NMP/g	–	SMEWW9221B	03/10/12
Escherichia Coli	<3,0	UFC/g	–	Cont. em Placas	03/10/12
Bolores e Leveduras em 25g	<1,0 X 10	UFC/g	–	Cont. em Placas	03/10/12
Salmonellas	Ausência	em 25g	–	AOACC 967.26	03/10/12

A Custódia das amostras é de 15 dias após emissão do relatório de ensaios, exceto para solos que é 90 dias e água que é 2 dias. Não se aplica a amostras perecíveis. Os resultados têm significado restrito e aplicam-se somente às amostras ensaiadas. Este relatório somente poderá ser reproduzido em sua totalidade. O ITPS se isenta de qualquer responsabilidade pela reprodução parcial do mesmo.

INSTITUTO TECNOLÓGICO E DE PESQUISAS DO ESTADO DE SERGIPE

Rua Campo do Brito, Nº371, Treze de Julho, CEP 49.020-380 Aracaju - SE - Brasil
Fone (79) 3179-8081/8087 Fax (79) 3179-8087/8090
CNPJ 07.258.529/0001-59

Relatório de Ensaios ITPS Nº 3483/12

Revisão 03

Cliente	LUCIANA COELHO MENDONÇA	**Telefone**	
Endereço	RUA FRANCISCO RABELO LEITE NETO, 990 CASA 8	**Contato(s)**	8817-9580
e-mail	lumendon@uol.com.br	**Fax**	
Amostra(s)	Rações	**Recepção**	03/10/12

Motivo da revisão 03: retirar a analise de contagem.

Legenda
NMP: Número Mais Provável.
SMEWW: Standard Methods for the Examination of Water and Wastewater, 21ª. ed. Washington, 2005.
UFC: Unidade formadora de colônia.
LQ: Limite de Quantificação do Método.

Informações de Coleta

Coleta efetuada pelo cliente.
A descrição do material ensaiado é de inteira responsabilidade do cliente.
****
PARTE AÉREA MOÍDA DE GIRASSOL
****

Aracaju,04 de abril de 2013

Angela de Menezes
Barreto
Quimico
CRQ-SE - 07200183
Lab. Microbiologia

Documento verificado e aprovado por meios eletrônicos

MIX
Papier aus verantwortungsvollen Quellen
Paper from responsible sources
FSC® C105338

FSC
www.fsc.org

Printed by Books on Demand GmbH, Norderstedt / Germany